ORDER NUMBER EA-FAA-T-8080-10BX

AVIATION MECHANIC
GENERAL

W9-DEW-238

QUESTION BOOK
INCLUDING ANSWERS, EXPLANATIONS AND REFERENCES

1988

International Standard Book Number 0-89100-326-6
For sale by: IAP, Inc.
Mail To: P.O. Box 10000, Casper, WY 82602-1000
Ship To: 7383 6WN Road, Casper, WY 82604-1835
(800) 443-9250 • (307) 266-3838 • FAX: 307-472-5106

IAP, Inc.
7383 6WN Road, Casper, WY 82604-1835

Answers, Explanations and References
Copyright 1988 by IAP, Inc.
All Rights Reserved
Printed in the United States of America

CONTENTS

PREFACE

This question book has been developed by the FAA (Federal Aviation Administration) for use by applicants who are preparing for the Aviation Mechanic General Written Test and also for use by FAA Testing Centers and FAA Designated Written Test Examiners in administering the test. It is issued as FAA–T–8080–10A, Aviation Mechanic General Question Book, and is available to the public from:

> Superintendent of Documents
> U.S. Government Printing Office
> Washington, D.C.   20402

or from U.S. Government Printing Office Bookstores located in major cities throughout the U.S.

The questions included in this publication are predicated on regulations, principles, and practices that were valid at the time of publication. The question selection sheets prepared for use with this question book are security items and are revised at frequent intervals.

The Federal Aviation Administration does not supply the correct answers to questions included in this book. Students should determine the answers by research and study, by working with instructors, or by attending ground schools. The Federal Aviation Administration is NOT responsible for either the content of commercial reprints of this book or the accuracy of the answers they may list.

Comments regarding this publication should be directed to:

> U.S. Department of Transportation
> Federal Aviation Administration
> Aviation Standards National Field Office
> Examinations Standards Branch
> Airworthiness Section, AVN–133
> P.O. Box 25082
> Oklahoma City, Oklahoma   73125

REFERENCE GUIDE

EA-ITP-GB . . . . AVIATION TECHNICIAN INTEGRATED TRAINING PROGRAM, GENERAL SECTION TEXTBOOK, IAP, INC.

EA-ITP-AB . . . . AVIATION TECHNICIAN INTEGRATED TRAINING PROGRAM, AIRFRAME SECTION TEXTBOOK, IAP, INC.

EA-ITP-P . . . . AVIATION TECHNICIAN INTEGRATED TRAINING PROGRAM, POWERPLANT SECTION TEXTBOOK, IAP, INC.

AIRCRAFT ELECTRICITY AND ELECTRONICS, THIRD EDITION, McGRAW HILL

AIRCRAFT MAINTENANCE AND REPAIR, FOURTH EDITION, McGRAW HILL

AIRCRAFT POWERPLANTS, FOURTH EDITION, McGRAW HILL

POWERPLANTS FOR AEROSPACE VEHICLES, THIRD EDITION, McGRAW HILL

AIRCRAFT GAS TURBINE ENGINE TECHNOLOGY, McGRAW HILL

AIRCRAFT PROPULSION POWERPLANTS, EDUCATIONAL PUBLISHERS, INC.

THE AVIATION MECHANIC'S MANUAL, McGRAW HILL

AC39-7A . . . . AIRWORTHINESS DIRECTIVES FOR GENERAL AVIATION AIRCRAFT

AC43-4 . . . . FAA ADVISORY CIRCULAR, CORROSION CONTROL

AC43-9A . . . . MAINTENANCE RECORDS: GENERAL AVIATION AIRCRAFT

AC43.13-1A . . . . ACCEPTABLE METHODS, TECHNIQUES, AND PRACTICES, AIRCRAFT INSPECTION AND REPAIR

AC43.13-2A . . . . AIRCRAFT ALTERATIONS

EA-AC 61-13B . . . . BASIC HELICOPTER HANDBOOK, U.S. DEPT. OF TRANSPORTATION

AC65-9A . . . . AIRFRAME AND POWERPLANT MECHANICS GENERAL HANDBOOK, U.S. DEPT. OF TRANSPORTATION

AC65-12 . . . . AIRFRAME AND POWERPLANT MECHANICS POWERPLANT HANDBOOK, U.S. DEPT. OF TRANSPORTATION

AC65-12A . . . . AIRFRAME AND POWERPLANT MECHANICS POWERPLANT HANDBOOK, U.S. DEPT. OF TRANSPORTATION

AC65-15A . . . . AIRFRAME AND POWERPLANT MECHANICS AIRFRAME HANDBOOK, U.S. DEPT. OF TRANSPORTATION

AC65-19B . . . . INSPECTION AUTHORIZATION STUDY GUIDE

EA-FAR . . . . FEDERAL AVIATION REGULATIONS, HANDBOOK FOR AVIATION MECHANICS, IAP, INC.

EA-AAC-1 . . . . AIRCRAFT AIRCONDITIONING (VAPOR CYCLE), IAP, INC.

EA-AGV . . . . AIRCRAFT GOVERNORS, IAP, INC.

EA-AH-1 . . . . AIRCRAFT HYDRAULIC SYSTEMS, IAP, INC.

EA-AOS-1 . . . . AIRCRAFT OXYGEN SYSTEMS, IAP, INC.

EA-APC . . . . AIRCRAFT PROPELLERS AND CONTROLS, IAP, INC.

EA-FMS-1 . . . . AIRCRAFT FUEL METERING SYSTEMS, IAP, INC.

EA-TEP . . . . AIRCRAFT GAS TURBINE POWERPLANTS, IAP, INC.

EA-ATD-2 . . . . AIRCRAFT TECHNICAL DICTIONARY, SECOND EDITION IAP, INC.

MAXIMUM TIME ALLOWED FOR TEST: 2 HOURS

Materials to be used with this question book when used for airman testing:

1. Airman Written Test Application which includes the answer sheet.
2. Question Selection Sheet which identifies the questions to be answered.
3. Plastic overlay sheet which can be placed over electrical drawings, graphs, and charts for making pencil marks or for tracing schematics for analysis purposes. If you are not provided the overlay, request one from the test monitor.

GENERAL INSTRUCTIONS

1. Read the instructions on page 1 of the Airman Written Test Application and complete the form on page 4.

2. The question numbers in this question book are numbered consecutively beginning with No. 4001. Refer to the question selection sheet to determine which questions to answer.

3. For each item on the answer sheet, find the appropriate question in the book.

4. Mark your answer in the space provided for that item on the answer sheet.

5. Remember:

   Read each question carefully and avoid hasty assumptions. Do not answer until you understand the question. Do not spend too much time on any one question. Answer all of the questions that you readily know and then reconsider those you find difficult.

   If a regulation is changed after this question book is printed, you will receive credit until the affected questions are revised.

The minimum passing grade is 70.

WARNING

§ 65.18 Written tests: cheating or other un-authorized conduct.

(a) Except as authorized by the Administrator, no person may—

   (1) Copy, or intentionally remove, a written test under this Part;

   (2) Give to another, or receive from another, any part or copy of that test;

   (3) Give help on that test to, or receive help on that test from, any person during the period that test is being given;

   (4) Take any part of that test in behalf of another person;

(5) Use any material or aid during the period that test is being given; or

(6) Intentionally cause, assist, or participate in any act prohibited by this paragraph.

(b) No person who commits an act prohibited by paragraph (a) of this section is eligible for any airman or ground instructor certificate or rating under this chapter for a period of one year after the date of that act. In addition, the commission of that act is a basis for suspending or revoking any airman or ground instructor certificate or rating held by that person.

INTRODUCTION

The requirements for a mechanic certificate and ratings, and the privileges, limitations, and general operating rules for certificated mechanics are prescribed in Federal Aviation Regulations Part 65, Certification: Airmen Other Than Flight Crewmembers. Any person who applies and meets the requirements is entitled to a mechanic certificate.

At FAA Testing Centers, or at an FAA Designated Written Test Examiner's facility, the applicant is issued a "clean copy" of this question book, an appropriate 50-item question selection sheet which indicates the specific questions to be answered, and AC Form 8080-3, Airman Written Test Application, which contains the answer sheet. The question book contains all the supplementary material required to answer the questions. Supplementary material, such as an illustration, will normally be found within one page of the question with which it is associated. Where this is not practicable, page reference numbers will be given.

THE WRITTEN TEST

Questions and Scoring

The test questions are of the multiple-choice type. Answers to questions listed on the question selection sheet should be marked on the answer sheet of AC Form 8080-3, Airman Written Test Application. Directions should be read carefully before beginning the test. Incomplete or erroneous personal information entered on this form delays the scoring process.

The answer sheet is sent to the Mike Monroney Aeronautical Center in Oklahoma City where it is scored by a computer. Shortly thereafter AC Form 8080-2, Airman Written Test Report, is sent to the applicant listing the score and test questions missed.

The applicant must present this report for an oral and practical test, or for retesting in the event of written test failure.

Taking the Test

The test may be taken at FAA Testing Centers, FAA Written Test Examiners' facilities, or other designated places. After completing the test, the applicant must surrender the issued question book, question selection sheet, answer sheet, and any papers used for computations or notations, to the monitor before leaving the test room.

When taking the test, the applicant should keep the following points in mind:

1. Answer each question in accordance with the latest regulations and procedures.
2. Read each question carefully before looking at the possible answers. You should clearly understand the problem before attempting to solve it.
3. After formulating an answer, determine which of the alternatives most nearly corresponds with that answer. The answer chosen should completely resolve the problem.
4. From the answers given, it may appear that there is more than one possible answer; however, there is only one answer that is correct and complete. The other answers are either incomplete or are derived from popular misconceptions.
5. If a certain question is difficult for you, it is best to proceed to other questions. After the less difficult questions have been answered, return to those which gave you difficulty. Be sure to indicate on the question selection sheet the questions to which you wish to return.
6. When solving a computer problem, select the answer nearest your solution. The problem has been checked with various types of computers; therefore, if you have solved it correctly, your answer will be closer to the correct answer than to any of the other choices.
7. To aid in scoring, enter personal data in the appropriate spaces on the test answer sheet in a complete and legible manner. Enter the test number printed on the question selection sheet.

Retesting—FAR 65.19

Applicants who receive a failing grade may apply for retesting by presenting their AC Form 8080-2, Airman Written Test Report—

(1) after 30 days from the date the applicant failed the test; or
(2) the applicant may apply for retesting before the 30 days have expired upon presenting a signed statement from an airman holding the certificate and rating sought by the applicant, certifying that the airman has given the applicant additional instruction in each of the subjects failed and that the airman considers the applicant ready for retesting.

NOTE: Blank spaces may appear on several pages of this publication. This permits referenced figures and tables to appear as close as possible to their related questions.

AVIATION MECHANIC GENERAL

5001. The working voltage of a capacitor to which a.c. or pulsating d.c. is applied should be

1—the same as or greater than the applied voltage.
2—at least 50 percent greater than the applied voltage.
3—1.41 times the applied voltage.
4—.707 times the applied voltage.

5002. If a circuit contains 10 ohms of resistance, 20 ohms of inductive reactance, and 30 ohms of capacitive reactance, it is said to be

1—inductive.
2—in resonance.
3—resistive.
4—capacitive.

5003. The opposition offered by a coil to the flow of alternating current is called

1—conductivity.
2—impedance.
3—reluctance.
4—inductive reactance.

5004. An increase in which of the following factors will cause an increase in the inductive reactance of a circuit?

1—Inductance and frequency.
2—Capacitance and voltage.
3—Resistance and voltage.
4—Resistance and capacitive reactance.

5005. The resistive force in a d.c. electrical circuit is usually measured in ohms and referred to as

1—resistance.
2—capacitance.
3—reactance.
4—inductance.

5006. When the capacitive reactance in an a.c. electrical circuit is equal to the inductive reactance, the circuit is said to be

1—in correct voltage phase angle.
2—in correct current phase angle.
3—out of phase.
4—resonant.

5007. In an alternating current circuit, the effective voltage

1—is equal to the maximum instantaneous voltage.
2—is greater than the maximum instantaneous voltage.
3—may be greater than or less than the maximum instantaneous voltage.
4—is less than the maximum instantaneous voltage.

5008. The amount of electricity a capacitor can store is directly proportional to

1—the distance between the plates and inversely proportional to the plate area.
2—the plate area and is not affected by the distance between the plates.
3—the plate area and inversely proportional to the distance between the plates.
4—the distance between the plates and is not affected by the plate area.

5009. A transformer with a step–up ratio of 5 to 1 has a primary voltage of 24 volts and a secondary amperage of 0.20 ampere. What is the primary amperage (disregard losses)?

1—1 ampere.
2—4.8 amperes.
3—0.40 ampere.
4—Cannot be determined from the information given.

5010. What is the phase relationship between the current and voltage in an inductive circuit?

1—The current lags the voltage by 0°.
2—The current lags the voltage by 90°.
3—The current leads the voltage by 90°.
4—The current leads the voltage by 0°.

5011. Current flow is measured in terms of

1—amperes.
2—volts.
3—watts.
4—electron flow.

5012. Unless otherwise specified, any values given for current or voltage in an alternating current circuit are assumed to be

1—average values.
2—instantaneous values.
3—effective values.
4—maximum values.

5013. Which of the following will require the most electrical power during operation?

(Note: 1 hp. = 746 watts.)

1—A 12–volt motor requiring 8 amperes.
2—Four 30–watt lamps in a 12–volt parallel circuit.
3—Two lights requiring 3 amperes each in a 24–volt parallel system.
4—A 1/10–horsepower, 24–volt motor which is 75 percent efficient.

5014. How much power must a 24-volt generator furnish to a system which contains the following loads?

UNIT	RATING
One motor (75 percent efficient)............................1/5 hp.	
Three position lights....................................20 watts each	
One heating element..5 amp.	
One anticollision light...3 amp.	

(Note: 1 hp. = 746 watts.)

1—18.75 watts.
2—402 watts.
3—385 watts.
4—450 watts.

5015. How many amperes will be required by a 24-volt, 1/3-horsepower electric motor, when operating at its rated load?

(Note: 1 hp. = 746 watts.)

1—10.4.
2—13.8.
3—7.9.
4—25.6.

5016. If a unit in a 28-volt aircraft electrical system has a resistance of 10 ohms, how much power will it use?

1—280 watts.
2—7.84 watts.
3—78.4 watts.
4—28 watts.

5017. A 12-volt electric motor has 1,000 watts input and 1 hp. output. Maintaining the same efficiency, how much input power will a 24-volt, 1-hp. electric motor require?

(Note: 1 hp. = 746 watts.)

1—1,000 watts.
2—2,000 watts.
3—500 watts.
4—Cannot be determined from the information given.

5018. How many amperes will a 28-volt generator be required to supply to a circuit containing five lamps in parallel, three of which have a resistance of 6 ohms each and two of which have a resistance of 5 ohms each?

1—1.11 amperes.
2—1 ampere.
3—0.9 ampere.
4—25.23 amperes.

5019. Which of the following is the rate of doing work equal to 1 hp.?

1—33,000 ft.–lb. per minute.
2—746 ft.–lb. per second.
3—3,300 ft.–lb. per minute.
4—55 ft.–lb. per second.

5020. The wattage rating of a carbon resistor is determined by

1—a gold band.
2—a silver band.
3—the size of the resistor.
4—a red band.

5021. The potential difference between two conductors which are insulated from each other is measured in

1—ohms.
2—volts.
3—amperes.
4—coulombs.

5022. The ratio of the true power to the apparent power in an a.c. electrical circuit is called the power factor. If the true power and the power factor of a circuit are known, the apparent power can be determined by

1—multiplying the true power times 100 times the power factor.
2—multiplying the power factor times 100 times the true power.
3—dividing the true power times 100 by the power factor.
4—dividing the power factor times 100 by the true power.

5023. A 24-volt source is required to furnish 48 watts to a parallel circuit consisting of four resistors of equal value. What is the voltage drop across each resistor?

1—12 volts.
2—6 volts.
3—3 volts.
4—24 volts.

5024. When calculating power in a reactive or inductive a.c. circuit, the true power is

1—more than the apparent power.
2—more than the apparent power in a reactive circuit and less than the apparent power in an inductive circuit.
3—less than the apparent power in a reactive circuit and more than the apparent power in an inductive circuit.
4—less than the apparent power.

5025. Determine the power furnished in watts by the generator of the circuit in Figure 1.

1—288 watts.
2—24 watts.
3—48 watts.
4—12 watts.

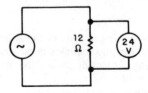

Figure 1

5026. If resistor R_5 is disconnected at the junction of R_4 and R_3 as shown, what will the ohmmeter read? (See Figure 2.)

1—9 ohms.
2—2.76 ohms.
3—3 ohms.
4—12 ohms.

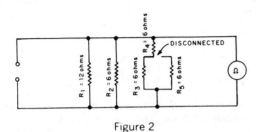

Figure 2

5027. Which of the following electrical measuring instruments is most likely to contain its own source of electrical power?

1—Wattmeter.
2—Ammeter.
3—Voltmeter.
4—Ohmmeter.

5028. The D'Arsonval—type meter movement used in an ammeter, voltmeter, or ohmmeter measures

1—current flow through the movement.
2—potential difference across the movement.
3—amount of resistance in series with the movement.
4—electrical power consumed by the movement.

5029. If resistor R_3 is disconnected at terminal D, what will the ohmmeter read? (See Figure 3.)

1—Infinite resistance.
2—0 ohm.
3—10 ohms.
4—20 ohms.

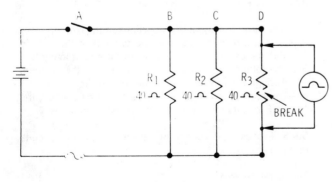

Figure 3

5030. With an ohmmeter connected into the circuit as shown, what will the ohmmeter read in Figure 4?

1—20 ohms.
2—Infinite resistance.
3—0 ohm.
4—10 ohms.

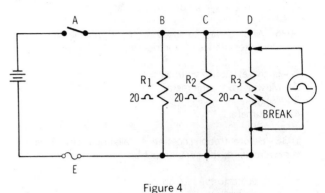

Figure 4

5031. In Figure 5, how many instruments (voltmeters and ammeters) are installed correctly?

1—Three.
2—One.
3—Two.
4—Four.

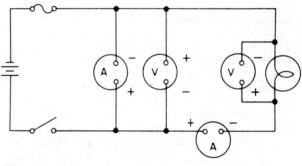

Figure 5

5032. The secondary voltage of a transformer depends upon the efficiency of the transformer and the ratio of the number of turns in the primary winding to the

1—number of turns in the secondary winding.
2—amount of current flowing in the primary winding.
3—material from which the core is constructed.
4—distance separating the windings.

5033. The correct way to connect a test voltmeter in a circuit is

1—in series with a unit.
2—between source voltage and the load.
3—in parallel with a unit.
4—to place one lead on either side of the fuse.

5034. A cabin–entry light of 10 watts and a dome light of 20 watts are connected in parallel to a 30–volt source. If the voltage across the 10–watt light is measured, it will be

1—one–third of the input voltage.
2—twice the voltage across the 20–watt light.
3—equal to the voltage across the 20–watt light.
4—half the voltage across the 20–watt light.

5035. What device is used to measure the very high insulation resistance of electric cables?

1—High–resistance voltmeter.
2—Moving iron–vane meter.
3—Megger.
4—Multimeter.

5036. Before troubleshooting an electrical circuit with a continuity light, the circuit must be

1—connected to the aircraft battery.
2—connected to the aircraft generator.
3—isolated.
4—connected to an external source of power.

5037. A 14–ohm resistor is to be installed in a series circuit carrying .05 ampere. How much power will the resistor be required to dissipate?

1—At least .70 milliwatt.
2—At least 35 milliwatts.
3—Less than .035 watt.
4—Less than .70 milliwatt.

5038. What is the maximum number of electrical wire terminals that can be installed on one stud?

1—Four terminals per stud.
2—Three terminals per stud.
3—Two terminals per stud.
4—As many terminals as you can stack on and still have the required number of threads showing through the nut.

5039. What would be the measured voltage of the series circuit in Figure 6 between terminals A and B?

1—1.5V.
2—3.0V.
3—4.5V.
4—6.0V.

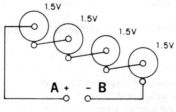

Figure 6

5040. The efficiency of power in an a.c. circuit is expressed by what term?

1—Volt–amperes.
2—True power.
3—Power factor.
4—Apparent power.

5041. The current in a 60–watt, 120–volt electric light bulb is

1—0.8 amp.
2—2 amps.
3—1/3 amp.
4—1/2 amp.

5042. Which of the following will require the most electrical power? (Note: 1 hp. = 746 watts.)

1—Four 30–watt lamps arranged in a 12–volt parallel circuit.
2—A 12–volt landing gear retraction motor which requires 8 amperes when operating the landing gear.
3—A 1/10–horsepower, 24–volt motor which is 75 percent efficient.
4—A 24–volt anticollision light circuit consisting of two light assemblies which require 3 amperes each during operation.

5043. What unit is used to express electrical power?

1—Coulomb.
2—Volt.
3—Watt.
4—Ampere.

5044. What is the operating resistance of a 30–watt light bulb designed for a 28–volt system?

1—30 ohms.
2—1.07 ohms.
3—26 ohms.
4—0.93 ohm.

5045. Which of the following statements is correct when made in reference to a parallel circuit?

1—The current is equal in all portions of the circuit.
2—The current in amperes is the product of the EMF in volts times the total resistance of the circuit in ohms.
3—The total current is equal to the sum of the currents through the individual branches of the circuit.
4—The current in amperes can be found by dividing the EMF in volts by the sum of the resistors in ohms.

5046. Diodes are used in electrical power circuits primarily as

1—current eliminators.
2—circuit cutout switches.
3—rectifiers.
4—power transducer relays.

5047. If three resistors of 3 ohms, 5 ohms, and 22 ohms are connected in series in a 28–volt circuit, how much current will flow through the 3–ohm resistor?

1—9.3 amperes.
2—1.05 amperes.
3—1.03 amperes.
4—0.93 ampere.

5048. A good conductor of electricity is a material

1—through or along which electrons move freely.
2—whose protons are all on the outside.
3—that contains few electrons.
4—through or along which protons move freely.

5049. A circuit has an applied voltage of 30 volts and a load consisting of a 10–ohm resistor in series with a 20–ohm resistor. What is the voltage drop across the 10–ohm resistor?

1—15 volts.
2—10 volts.
3—20 volts.
4—30 volts.

5050. Find the total current flowing in the wire between points C and D in Figure 7.

1—6.0 amperes.
2—2.4 amperes.
3—3.0 amperes.
4—0.6 ampere.

5051. Find the voltage across the 8–ohm resistor using the drawing in Figure 7.

1—2.4 volts.
2—12 volts.
3—20.4 volts.
4—24 volts.

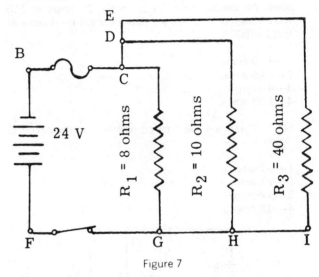

Figure 7

5052. Find the total resistance of the circuit in Figure 8.

1—16 ohms.
2—10.4 ohms.
3—2.6 ohms.
4—21.2 ohms.

$$R_t = R_c + R_1$$

$$R_a = \frac{1}{1/R_4 + 1/R_5}$$

$$R_b = R_a + R_2$$

$$R_c = \frac{1}{1/R_b + 1/R_3}$$

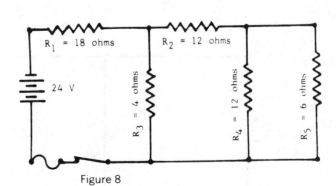

Figure 8

5

5053. Which of the following is correct in reference to electrical resistance?

1—Two electrical devices will have the same combined resistance if they are connected in series as they will have if connected in parallel.
2—If one of three bulbs in a parallel lighting circuit is removed, the total resistance of the circuit will become greater.
3—An electrical device that has a high resistance will use more power than one with a low resistance with the same applied voltage.
4—A 5–ohm resistor in a 12–volt circuit will use less current than a 10–ohm resistor in a 24–volt circuit.

5054. An electric cabin heater draws 25 amps at 110 volts. How much current will flow if the voltage is reduced to 85 volts?

1—19.3 amps.
2—44.0 amps.
3—4.4 amps.
4—1.93 amps.

5055 Determine the total current flow in the circuit in Figure 9.

1—0.2 amp.
2—1.4 amps.
3—0.4 amp.
4—0.8 amp.

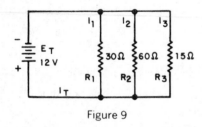

Figure 9

5056. The total resistance of the circuit in Figure 10 is

1—25 ohms.
2—35 ohms.
3—37 ohms.
4—17 ohms.

5057. Through which of the following will magnetic lines of force pass the most readily?

1—Copper.
2—Iron.
3—Aluminum.
4—Titanium.

5058. A 48–volt source is required to furnish 192 watts to a parallel circuit consisting of three resistors of equal value. What is the value of each resistor?

1—36 ohms.
2—4 ohms.
3—8 ohms.
4—12 ohms.

5059. Which is correct concerning a parallel circuit?

1—Total resistance will be smaller than the smallest resistor.
2—Total resistance will decrease when one of the resistances is removed.
3—Total voltage drop is the same as the total resistance.
4—Total amperage remains the same, regardless of the resistance.

5060. The voltage drop in a conductor of known resistance is dependent on

1—the voltage of the circuit.
2—the amount and thickness of wire insulation.
3—only the resistance of the conductor and does not change with a change in either voltage or amperage.
4—the amperage of the circuit.

5061. An electric motor malfunctions causing it to overheat, which will cause an incorporated thermal switch to

1—prevent an open circuit.
2—break the circuit.
3—close the circuit.
4—break the circuit when cooled.

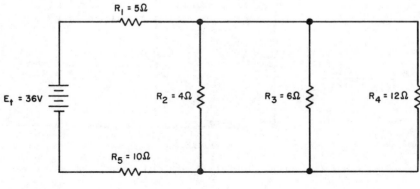

Figure 10

6

5062. With the landing gear retracted (see Figure 11), the red indicator light will not come on if an open occurs in wire

1—No. 19.
2—No. 7.
3—No. 16.
4—No. 17.

5063. Using Figure 11, the No. 7 wire is used to

1—open the DOWN indicator light circuit when the landing gear is retracted.
2—complete the PUSH–TO–TEST circuit.
3—open the UP indicator light circuit when the landing gear is retracted.
4—close the UP indicator light circuit when the landing gear is retracted.

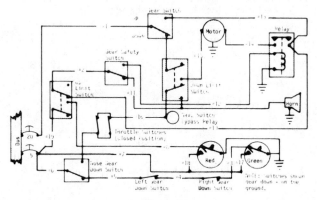

Figure 11

7

5064. What would be the effect if the PCO relay in Figure 12 failed to operate when the left–hand tank was selected?

1—The fuel pressure crossfeed valve would not open.
2—The fuel tank crossfeed valve would open.
3—The fuel tank crossfeed valve open light would illuminate.
4—The fuel pressure crossfeed valve open light would not illuminate.

5065. The TCO relay in Figure 12 will operate if 24 VDC is applied to the bus and the fuel tank selector is in the

1—right–hand tank position.
2—crossfeed position.
3—left–hand tank position.
4—normal position.

5066. With power to the bus and the fuel selector switched to the right–hand tank (see Figure 12), how many relays in the system are operating?

1—Three.
2—One.
3—Two.
4—Four.

5067. When electrical power is applied to the bus (see Figure 12), which relays are energized?

1—PCO and TCC.
2—PCC and TCC.
3—TCC and TCO.
4—PCO and PCC.

5068. Energize the circuit with the fuel tank selector switch selected to the left–hand position. Using the schematic in Figure 12, identify the switches that will change position.

1—5, 11, 12, 13, 15, 9, 10.
2—5, 6, 3, 7, 11, 13.
3—5, 6, 11, 15, 12, 13, 16.
4—5, 7, 11, 15.

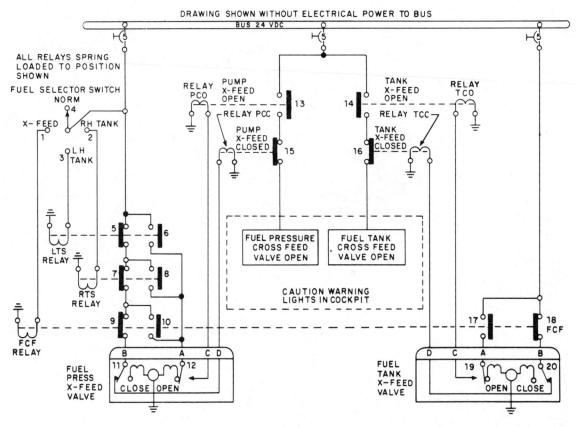

Figure 12

5069. Which of the components in Figure 13 is a potentiometer?

1—D.
2—E.
3—C.
4—M.

5070. In Figure 13, what electrical symbol is represented at E?

1—Fixed capacitor.
2—Fixed resistor.
3—Variable resistor.
4—Variable capacitor.

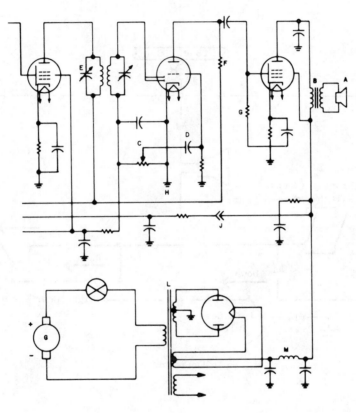

Figure 13

5071. When the landing gears are up and the throttles are retarded (see Figure 14), the warning horn will not sound if an open occurs in wire

1—No. 4.
2—No. 2.
3—No. 9.
4—No. 10.

5072. The control valve switch should be placed in the neutral position when the landing gears are down (see Figure 14) to

1—permit the test circuit to operate.
2—provide a ground for the red light.
3—prevent the warning horn from sounding when the throttles are closed.
4—remove the ground from the green light.

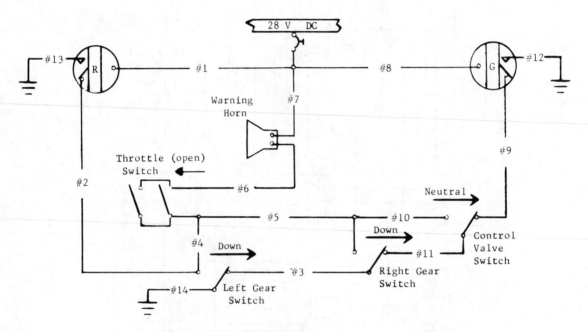

Figure 14

5073. Using Figure 15, under which condition will a ground be provided for the warning horn through both gear switches when the throttles are closed?

1—Right gear up and left gear down.
2—Anytime the gears malfunction.
3—Both gears up and the control valve out of neutral.
4—Left gear up and right gear down.

5074. When the throttles are retarded with only the right gear down (see Figure 15), the warning horn will not sound if an open occurs in wire

1—No. 5.
2—No. 13.
3—No. 8.
4—No. 6.

5075. When the landing gears are up and the throttles are retarded (see Figure 15), the warning horn will not sound if an open occurs in wire

1—No. 5.
2—No. 7.
3—No. 13.
4—No. 6.

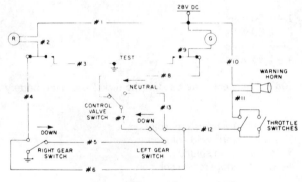

Figure 15

5076. Schematic diagrams indicate the location of individual components in the aircraft

1—with aircraft station numbers of each component.
2—on the title block by federal stock number.
3—with respect to each other within the system.
4—with detail drawings of each component.

5077. When referring to an electrical circuit diagram, what point is considered to be at zero voltage?

1—The ground reference.
2—The current limiter.
3—The fuse.
4—The switch.

5078. Troubleshooting an open circuit with a voltmeter as shown in this circuit (see Figure 16) will

1—permit current to flow and illuminate the lamp.
2—create a low resistance path and the current flow will be greater than normal.
3—restrict current flow and no voltage will appear on the voltmeter.
4—permit the battery voltage to appear on the voltmeter.

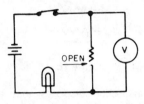

Figure 16

5079. A battery containing six lead–acid cells connected in series (2.1 volts per cell) has a terminal voltage of 10 volts when delivering 2 amperes to a load. What is the internal resistance of the battery?

1—1.3 ohms.
2—2.6 ohms.
3—5.0 ohms.
4—6.3 ohms.

5080. Using Figure 17, when the landing gear is down, the green light will not come on if an open occurs in wire

1—No. 7.
2—No. 6.
3—No. 16.
4—No. 17.

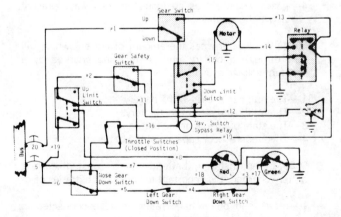

Figure 17

5081. Which symbol in Figure 18 represents a variable resistor?

1—C.
2—A.
3—B.
4—D.

A
B
C
D

Figure 18

5082. If an inspection discloses that a considerable amount of acid from a lead–acid battery has been spilled in the general area of the battery compartment, which of the following procedures should be followed?

1—Neutralize the spilled battery acid by applying sodium bicarbonate to the affected area followed by a water wash.
2—Apply sodium bicarbonate in powder form to the affected area.
3—Apply water to the affected area until the spilled battery acids turn a cloudy white.
4—Wipe the affected area with an oil–soaked cloth.

5083. A fully charged lead–acid battery will not freeze until extremely low temperatures are reached because

1—the acid is in the plates, thereby increasing the specific gravity of the solution.
2—most of the acid is in the solution.
3—the increased internal resistance generates sufficient heat to prevent freezing.
4—gases which act as an insulator are always present above the solution.

5084. What determines the amount of current which will flow through a battery while it is being charged by a constant voltage source?

1—The number of cells in the battery.
2—The total plate area of the battery.
3—The state of charge of the battery.
4—The ampere–hour capacity of the battery.

5085. When installing batteries on a constant–current battery charger, it is important to know that batteries of

1—more than one voltage rating may be connected in series and charged at the same time.
2—more than one voltage rating may be connected in parallel and charged at the same time.
3—only one voltage rating may be connected in series and charged at the same time.
4—only one voltage rating may be connected in parallel and charged at the same time.

5086. Aircraft batteries are usually rated according to voltage and ampere–hour capacity. A 35 ampere–hour battery rating based on the 5–hour discharge rate indicates

1—the battery will supply a current of 7 amperes for 5 hours.
2—the battery will not become discharged in less than 5 hours.
3—a battery discharge rate of 35 amperes must not be continued for more than 5 hours.
4—the battery will supply a current of 35 amperes for 5 hours.

5087. Some aircraft use more than one battery to increase the power available for starting and for operating emergency loads. How are the batteries connected so that the output power will increase but the voltage will remain the same?

1—By connecting the batteries to separate loads.
2—It is not possible.
3—By connecting the batteries in parallel.
4—By connecting the batteries in series.

5088. What is the rated voltage–per–cell of a nickel–cadmium battery?

1—1.30 volts.
2—2.00 volts.
3—2.25 volts.
4—1.55 volts.

5089. The electrolyte used in the nickel–cadmium battery is a

1—potassium hydroxide solution.
2—hydrochloric acid solution.
3—sulfuric acid solution.
4—potassium peroxide solution.

5090. Most aircraft storage batteries are rated according to

1—open–circuit voltage and closed–circuit voltage.
2—voltage and ampere hour capacity.
3—the maximum number of volt–amperes (power) the battery can furnish to a load.
4—battery voltage and volts per cell.

5091. (1) The electrolyte in a nickel–cadmium battery acts as a conductor.

(2) The electrolyte in a nickel–cadmium battery provides an insulator between the plates.

Regarding the above statements, which of the following is true?

1—Only No. 1 is true.
2—Only No. 2 is true.
3—Both No. 1 and No. 2 are true.
4—Neither No. 1 nor No. 2 is true.

5092. Which of the following is an indication of improperly torqued cell link connections of a nickel–cadmium battery?

1—Light spewing at the cell caps.
2—Low temperature in the cells.
3—Toxic and corrosive deposits of potassium carbonate crystals.
4—Heat or burn marks on the hardware.

5093. What is the most probable cause of an excessive amount of potassuim carbonate formation on a nickel–cadmium battery?

1—Battery overcharging.
2—Good battery ventilation.
3—Diluted electrolyte solution.
4—High internal resistance in the battery.

5094. When should water be added to a nickel–cadmium battery?

1—When the battery is in a discharged condition.
2—Before starting the charging cycle.
3—During the charging cycle.
4—Immediately after the charging cycle is completed.

5095. The electrolyte of a nickel–cadmium battery is the lowest when

1—the battery is being charged.
2—the battery is fully charged.
3—the battery is in a discharged condition.
4—the battery is under load condition.

5096. What is the cell voltage of a fully charged 19 cell, nickel–cadmium battery?

1—1.2 to 1.3 volts.
2—1.4 to 1.5 volts.
3—1.7 to 1.8 volts.
4—1.8 to 1.9 volts.

5097. Nickel–cadmium batteries which are stored for a long period of time will show a low fluid level because the

1—fluid evaporates through the vents.
2—battery is fully charged.
3—fluid level was not periodically replenished.
4—electrolyte becomes absorbed in the plates.

5098. How can the state-of-charge of a nickel–cadmium battery be determined?

1—By measuring the specific gravity of the electrolyte.
2—By a measured discharge.
3—By the temperature of the plates.
4—By the level of the electrolyte.

5099. Which of the following may result if water is added to a nickel–cadmium battery when it is not fully charged?

1—The cell temperatures will run too low for proper output.
2—The electrolyte will be absorbed by the plates during the charging cycle.
3—No adverse results since water may be added anytime.
4—Excessive spewing will occur during the charging cycle.

5100. Overcharging of nickel–cadmium batteries causes

1—excessive wetness around the cells, connectors, and cases.
2—consistently low electrolyte level.
3—formation of an excessive amount of potassium carbonate.
4—excessive plate sulfation.

5101. When a charging current is applied to a nickel–cadmium battery, the cells emit gas only

1—toward the end of the charging cycle.
2—at the start of the charging process.
3—when the electrolyte level is low.
4—if they are defective.

5102. What type of line is normally used in a mechanical drawing or blueprint to represent an edge or object not visible to the viewer?

1—Medium–weight dashed line.
2—Light solid line.
3—Alternate short and long heavy dashes.
4—Zigzag or wavy line.

5103. A line used to divide a drawing into equal or symmetrical parts is a

1—center line.
2—section line.
3—extension line.
4—cutting plane line.

5104. In the isometric view (see Figure 19) of a typical aileron balance weight, identify the view indicated by the arrow.

1—A.
2—B.
3—D.
4—C.

5105. Schematic diagrams

1—indicate the location of individual components in the aircraft.
2—do not give individual component location in the aircraft.
3—do not locate components within the system.
4—are used to locate parts and assemblies of parts for installation in aircraft.

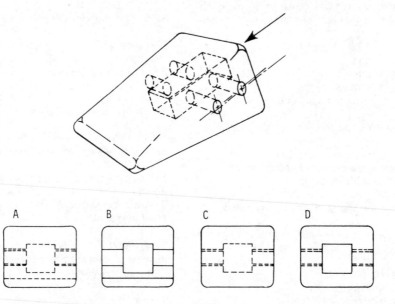

Figure 19

5106. (1) A detail drawing is a description of a single part.

(2) An assembly drawing is a description of an object made up of two or more parts.

Regarding the above statements, which of the following is true?

1—Only No. 1 is true.
2—Only No. 2 is true.
3—Neither No. 1 nor No. 2 is true.
4—Both No. 1 and No. 2 are true.

5107. Identify the bottom view of the object shown in Figure 20.

1—A.
2—C.
3—B.
4—D.

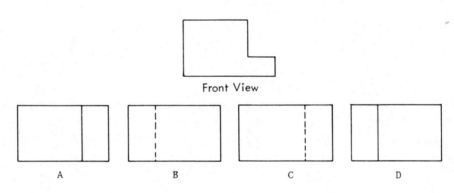

Front View

A   B   C   D

Figure 20

5108. What is a fuselage station number?

1—A measurement in inches from the datum or some other point chosen by the manufacturer.
2—A zone number used to locate a particular point.
3—A measurement in inches from the centerline or zero station of the aircraft.
4—A measurement in inches which always starts at the nose of the aircraft.

5109. What type of line is normally used in a mechanical drawing to represent the exposed surfaces of an object in sectional view?

1—Sectioning line.
2—Leader line.
3—Break line.
4—Outline or visible line.

5110. How many views are possible with an orthographic projection?

1—Four.
2—Six.
3—Three.
4—Five.

5111. In an orthographic projection, which of the following is true?

1—There are always at least two views.
2—It could have as many as eight views.
3—It must be accompanied with a pictorial drawing.
4—One–view, two–view, and three–view drawings are the most common.

5112. Identify the left side view of the object shown in Figure 21.

1—A.
2—B.
3—C.
4—D.

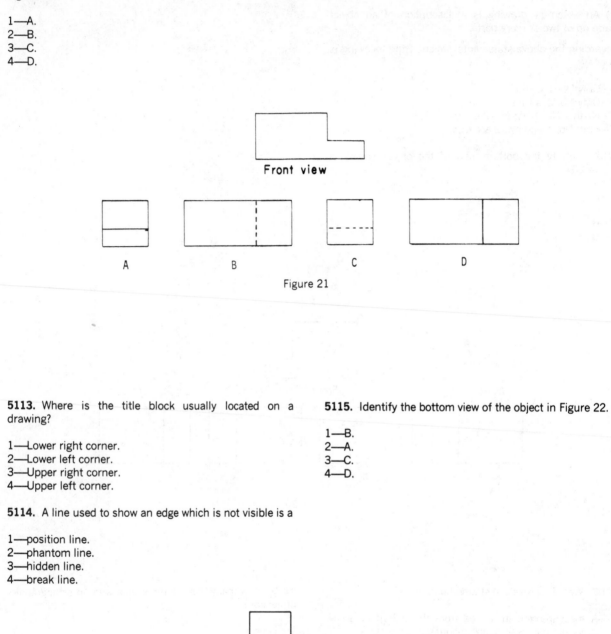

Front view

A   B   C   D

Figure 21

5113. Where is the title block usually located on a drawing?

1—Lower right corner.
2—Lower left corner.
3—Upper right corner.
4—Upper left corner.

5114. A line used to show an edge which is not visible is a

1—position line.
2—phantom line.
3—hidden line.
4—break line.

5115. Identify the bottom view of the object in Figure 22.

1—B.
2—A.
3—C.
4—D.

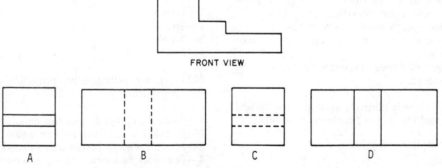

FRONT VIEW

A   B   C   D

Figure 22

16

5116. (1) Schematic diagrams indicate the location of individual components in the aircraft.

(2) Schematic diagrams indicate the location of components with respect to each other within the system.

Regarding the above statements, which of the following is true?

1—Only No. 1 is true.
2—Both No. 1 and No. 2 are true.
3—Only No. 2 is true.
4—Neither No. 1 nor No. 2 is true.

5117. Identify the phantom line used in aircraft drawings.

1—
2—
3—
4—

5118. Which of the following is used extensively in all aircraft maintenance and repair manuals and are invaluable in identifying and locating components and understanding the operation of various systems?

1—Drawing sketches.
2—Pictorial drawings.
3—Assembly drawings.
4—Installation diagrams.

5119. What are the proper procedural steps for sketching repairs and alterations? (See Figure 23.)

1—C, A, D, B.
2—B, C, A, D.
3—D, B, C, A.
4—A, C, D, B.

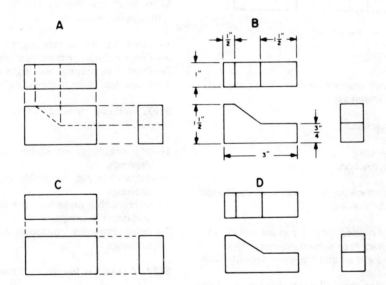

Figure 23

5120. What is the next step required for a working sketch of the illustration? (See Figure 24.)

1—Darken the object outlines.
2—Sketch extension and dimension lines.
3—Add notes, dimensions, title, and date.
4—Sketch at least two more views of the object.

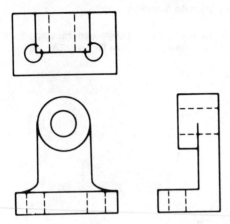

Figure 24

5121. When an aircraft drawing contains three separate views, the drawing is known as

1—separate view projection.
2—side–view drawing.
3—an isometric three–view.
4—an orthographic projection.

5122. Which of the following statements is applicable when using a sketch for making a part?

1—The sketch may be used only if supplemented with three–view orthographic projection drawings.
2—The sketch must show all information to manufacture the part.
3—The sketch need not show all necessary construction details.
4—A part is never made solely from a sketch.

5123. In orthographic projection drawings, it is often possible to portray an object clearly by the use of three views. When three–view projection is used, which of the following views are usually shown?

1—Top, front, and bottom.
2—Front, left side, and right side.
3—Top, front, and right side.
4—Front, back, and left side.

5124. What should be the first step of the procedure in sketching an aircraft wing skin repair?

1—Draw heavy guidelines.
2—Lay out the repair.
3—Draw the details.
4—Block in the views.

5125. (1) According to FAR Part 91, repairs to an aircraft skin should have a detailed dimensional sketch included in the permanent records.

(2) On occasion, a mechanic may need to make a simple sketch of a proposed repair to an aircraft, a new design, or a modification.

Regarding the above statements, which of the following is true?

1—Only No. 1 is true.
2—Only No. 2 is true.
3—Both No. 1 and No. 2 are true.
4—Neither No. 1 nor No. 2 is true.

5126. What are the six possible views of an object in orthographic projection?

1—Front, top, inside, rear, right side, and left side.
2—Front, outside, bottom, rear, right side, and left side.
3—Front, top, bottom, rear, right side, and left side.
4—Front, top, bottom, inside, right side, and left side.

5127. Working drawings may be divided into three classes. They are:

1—title drawings, installation drawings, and assembly drawings.
2—detail drawings, assembly drawings, and installation drawings.
3—orthographic projection drawings, pictorial drawings, and detail drawings.
4—detail drawings, pictorial drawings, and assembly drawings.

5128. A sketch is frequently drawn for use in

1—manufacturing a replacement part.
2—training of an airframe mechanic.
3—identifying the person drawing the sketch.
4—troubleshooting.

18

5129. What is the dimension of the chamfer in Figure 25?

1—0.0625 X 45°.
2—0.0625 R.
3—0.4062 R spherical.
4—0.5000 diameter.

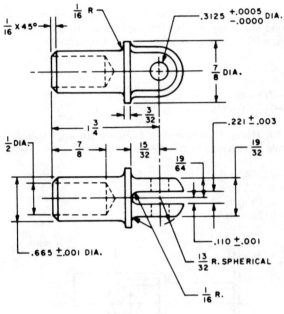

Figure 25

5130. Identify the bottom view of the object shown in Figure 26.

1—A.
2—B.
3—C.
4—D.

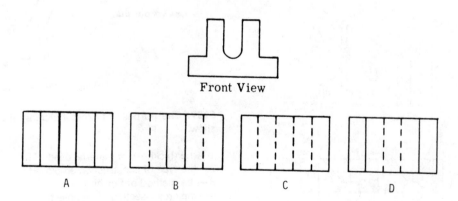

Front View

A   B   C   D

Figure 26

5131. What are the means of conveying measurements through the medium of drawings?

1—Dimensions.
2—Tolerances.
3—Edge distances.
4—Bend allowances.

5132. Identify the extension line in Figure 27.

1—C.
2—A.
3—B.
4—D.

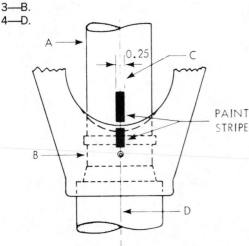

Figure 27

5133. The diameter of the holes in the finished object of Figure 28 is

1—3/4 inch.
2—31/64 inch.
3—1 inch.
4—1/2 inch.

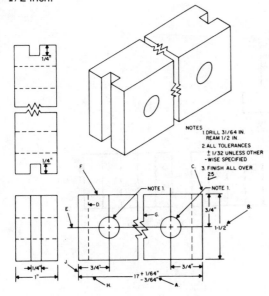

Figure 28

5134. What is the maximum diameter of the hole for the clevis pin in Figure 29?

1—0.500.
2—0.3125.
3—0.3130.
4—0.31255.

5135. What would be the minimum diameter of 4130 round stock required for the construction of the clevis in Figure 29 that would produce a machined surface?

1—55/64 inch.
2—1 inch.
3—7/8 inch.
4—1–1/16 inch.

5136. Using the information in Figure 29, what size drill would be required to drill the clevis bolthole?

1—21/64 inch.
2—19/64 inch.
3—5/16 inch.
4—9/32 inch.

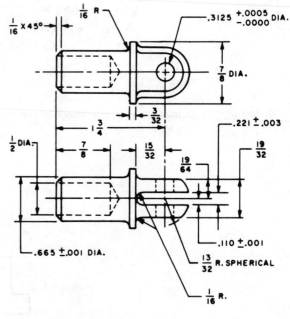

Figure 29

5137. The purpose of cross–hatching in a drawing is to show

1—the thickness of material used.
2—the type of construction employed.
3—the method of finishing.
4—the cross–section of an object.

5138. Zone numbers on aircraft blueprints are

1—used to locate parts, sections, and views on large drawings.
2—used to indicate different sections of the aircraft.
3—used to locate parts in the aircraft.
4—not used on working drawings.

5139. Schematic diagrams are used for

1—locating components in the aircraft.
2—troubleshooting.
3—identifying items in the aircraft.
4—manufacturing aircraft parts.

5140. When reading a blueprint, a dimension is given as 4.387″ + .005, − .002. Which of the following is true?

1—The maximum acceptable size is 4.389″.
2—The tolerance is .007.
3—The maximum acceptable size is 4.385″.
4—The minimum acceptable size is 4.387″.

5141. What is the allowable manufacturing tolerance for a bushing where the outside dimensions shown on the blueprint are: 1.0625 +.0025, −.0003?

1—.0028.
2—1.0650.
3—1.0647.
4—.0025.

5142. A hydraulic system schematic drawing would indicate the

1—specific location of the individual components within the aircraft.
2—direction of fluid flow through the system.
3—type and quantity of the hydraulic fluid.
4—part or model numbers of the individual components.

5143. What is the meaning of print tolerance as used in the title block of an aircraft blueprint?

1—Describes the size of the blueprint paper.
2—Shows proprietary rights of the blueprint.
3—Establishes the tolerance for a part when not otherwise shown by a dimension line.
4—Overrides any other dimension that may be shown on the blueprint.

5144. In Figure 30 the vertical distance between the top of the plate and the bottom of the lowest 15/64″ hole is

1—2.250.
2—2.242.
3—2.367.
4—3.312.

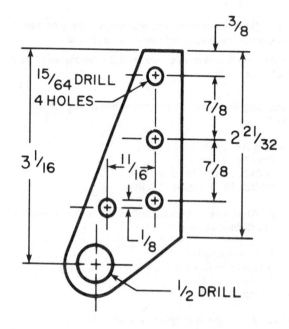

Figure 30

5145. (1) A measurement should not be scaled from an aircraft print because the paper shrinks or stretches when the print is made.

(2) When a detail drawing is made, it is carefully and accurately drawn to scale, and is dimensioned.

Regarding the above statements, which of the following is true?

1—Only No. 1 is true.
2—Only No. 2 is true.
3—Both No. 1 and No. 2 are true.
4—Neither No. 1 nor No. 2 is true.

5146. The drawings often used in illustrated parts manuals are

1—exploded–view drawings.
2—block drawings.
3—section drawings.
4—detail drawings.

5147. A drawing in which all of the parts are brought together as an assembly is called

1—a sectional drawing.
2—a detail drawing.
3—a block drawing.
4—an installation drawing.

5148. (1) An assembly drawing shows the way two or more parts are assembled into a complete unit.

(2) An installation drawing shows the relationship of all of the components.

Regarding the above statements which of the following is true?

1—Only No. 1 is true.
2—Only No. 2 is true.
3—Both No. 1 and No. 2 are true.
4—Neither No. 1 nor No. 2 is true.

5149. What type of drawing shows the wire size required for a particular installation?

1—A block diagram.
2—A schematic diagram.
3—A wiring diagram.
4—A pictorial diagram.

5150. What type of diagram is used to explain a principle of operation, rather than show the parts as they actually appear?

1—A pictorial diagram.
2—A schematic diagram.
3—A block diagram.
4—A wiring diagram.

5151. Determine the engine speed (RPM) necessary to develop 850 brake horsepower in a 2,000 cubic inch displacement aircraft reciprocating engine operating at 185 BMEP. (See Figure 31.)

1—1,800.
2—1,200.
3—1,400.
4—1,600.

5152. An aircraft reciprocating engine has 1,830 cubic inch displacement, develops 1,250 brake horsepower at 2,500 RPM. What is the brake mean effective pressure? (See Figure 31.)

1—217.
2—190.
3—205.
4—225.

5153. An aircraft reciprocating engine has 2,800 cubic inch displacement, develops 2,000 brake horsepower, and indicates 270 brake mean effective pressure. What is the engine speed (RPM)? (Use Figure 31.)

1—2,200.
2—2,100.
3—2,300.
4—2,400.

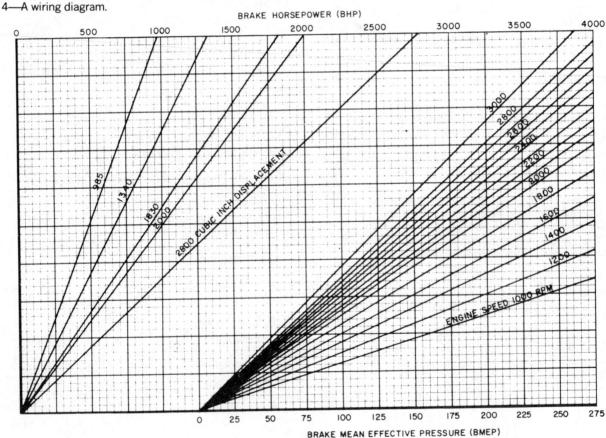

Figure 31

5154. Using Figure 32, determine the cable size of a 40–foot length of single cable in free air with a continuous rating, running from a bus to the equipment in a 28–volt system with a 15–ampere load and a 1–volt drop.

1—No. 10.
2—No. 11.
3—No. 18.
4—No. 6.

5155. Using Figure 32, determine the smallest size wire acceptable for carrying 80 amperes of continuous current using a single wire in free air.

1—No. 2.
2—No. 8.
3—No. 6.
4—No. 4.

5156. An aircraft reciprocating engine has 985 cubic inch displacement, develops 500 brake horsepower, and indicates 200 brake mean effective pressure. What is the engine speed (RPM)? (Use Figure 31.)

1—1600.
2—1800.
3—2000.
4—2100.

5157. Using Figure 33, determine the proper tension for a 1/8 inch cable (7 X 19) if the temperature is –30°F.

1—32 lbs.
2—40 lbs.
3—48 lbs.
4—56 lbs.

5158. An aircraft reciprocating engine has 2,800 cubic inch displacement, develops 2,000 brake horsepower at 2,200 RPM. What is the brake mean effective pressure. (Use Figure 31.)

1—257.
2—210.
3—242.
4—275.

5159. Using Figure 32, determine the maximum length of a No. 14 single wire in free air, running from a bus to the equipment in a 28–volt system with a 30–ampere continuous load and a 1–volt drop.

1—11 feet.
2—9 feet.
3—13 feet.
4—15 feet.

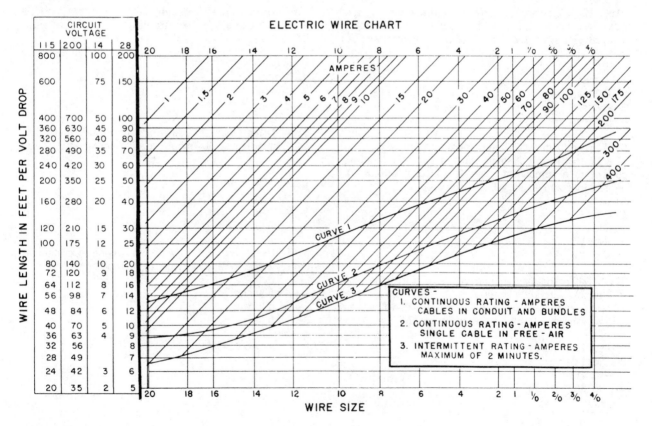

Figure 32

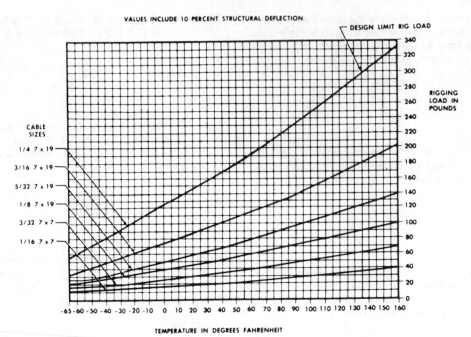

VALUES INCLUDE 10 PERCENT STRUCTURAL DEFLECTION

DESIGN LIMIT RIG LOAD

RIGGING LOAD IN POUNDS

CABLE SIZES

1/4 7 x 19
3/16 7 x 19
5/32 7 x 19
1/8 7 x 19
3/32 7 x 7
1/16 7 x 7

TEMPERATURE IN DEGREES FAHRENHEIT

Figure 33

5160. Using Figure 32, determine the maximum length of a No. 16 cable to be installed from a bus to the equipment in a 28–volt system with a 25 ampere intermittent load and a 1–volt drop.

1—8 feet.
2—10 feet.
3—12 feet.
4—14 feet.

5161. Using Figure 32, determine the minimum wire size of a 15–foot length of single cable in a bundle supplying current to a component rated at 14 volts, with a 20–ampere load and an allowable .5–volt drop.

1—No. 8.
2—No. 10.
3—No. 12.
4—No. 14.

5162. Determine the minimum wire size of a single cable in a bundle carrying a continuous current of 20 amperes 10 feet from the bus to the equipment in a 28–volt system with an allowable 1–volt drop. (See Figure 32.)

1—No. 12.
2—No. 14.
3—No. 16.
4—No. 18.

5163. Using Figure 33, determine the proper tension for a 3/16–inch cable (7 X 19 extra flex) if the temperature is 87° F.

1—132 lb.
2—140 lb.
3—136 lb.
4—144 lb.

5164. Using Figure 34, determine how much fuel would be required for a 30–minute reserve operating at 2,300 RPM.

1—25.7 lb.
2—22.3 lb.
3—44.6 lb.
4—49.8 lb.

5165. Using Figure 34, with the engine operating at cruise, 2,350 RPM, determine the fuel consumption.

1—49.5 lb./hr.
2—51.5 lb./hr.
3—55.5 lb./hr.
4—58.7 lb./hr.

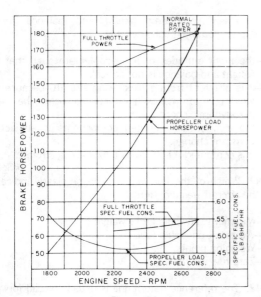

Figure 34

5166. Determine the maximum length of a No. 16 cable that can be used in a bundle supplying a 14 volt, 10 amps continuous load with maximum acceptable voltage drop of 1 volt between the bus and the utilizing equipment. (See Figure 32.)

1—10.5 feet.
2—20.5 feet.
3—30.5 feet.
4—40.5 feet.

5167. Determine the maximum length of a No. 12 single cable that can be used between a 28–volt bus and a component utilizing 20 amperes continuous load in free air with a maximum acceptable 1–volt drop. (See Figure 32.)

1—10.5 feet.
2—22.5 feet.
3—26.5 feet.
4—12.5 feet.

5168. Which of the following tasks are completed prior to weighing an aircraft to determine its empty weight?

1—Remove all items on the aircraft equipment list; drain fuel and hydraulic fluid.
2—Remove all items except those on the aircraft equipment list; drain fuel and hydraulic fluid.
3—Remove all items on the aircraft equipment list; drain fuel, compute oil and hydraulic fluid weight.
4—Remove all items except those on the aircraft equipment list; drain fuel and fill hydraulic reservoir.

5169. The useful load of an aircraft consists of the

1—crew, fuel, passengers, and cargo.
2—crew, fuel, oil, and fixed equipment.
3—crew, passengers, fuel, oil, cargo, and fixed equipment.
4—crew, powerplant, fuel, oil, cargo, and passengers.

5170. Before weighing an aircraft, it is necessary to become familiar with the aircraft empty weight and CG range in the weight and balance information in

1—the particular aircraft's weight and balance records.
2—Advisory Circular 43.13–1A, chapter 12.
3—the applicable Aircraft Specification or Type Certificate Data Sheet.
4—the manufacturer's service bulletins and letters.

5171. In the theory of weight and balance, what is the name of the distance from the fulcrum to an object?

1—Lever arm.
2—Beam length.
3—Balance arm.
4—Fulcrum arm.

5172. When preparing an aircraft for weighing, which of the following should be filled unless otherwise noted in the aircraft specifications or manufacturer's instructions?

1—Hydraulic reservoirs.
2—Lavatory tanks.
3—Wash water reservoirs.
4—Drinking water reservoirs.

5173. (1) Private aircraft are required by regulations to be weighed periodically.

(2) Private aircraft are required to be weighed after making any alteration.

Regarding the above statements, which of the following is true?

1—Neither No. 1 nor No. 2 is true.
2—Only No. 1 is true.
3—Only No. 2 is true.
4—Both No. 1 and No. 2 are true.

5174. What document will reference the required equipment needed to maintain validity of a standard airworthiness certificate?

1—Manufacturer's maintenance manual.
2—Advisory Circular 43.13–1A.
3—The aircraft's latest weight and balance information.
4—Aircraft Specification or Type Certificate Data Sheet.

5175. Which of the following designates the location of the reference points for leveling aircraft on the ground for weighing purposes?

1—Certificated A and P mechanic.
2—FAA Administrator.
3—Certificated repair station.
4—Aircraft manufacturer.

5176. To obtain useful weight data for purposes of determining the CG, it is necessary that an aircraft be weighed

1—with the main weighing points located within the normal CG limits.
2—in a level flight attitude.
3—with all items of useful load installed.
4—with at least minimum fuel (1/12-gallon per METO horsepower) in the fuel tanks.

5177. What unit of measurement is used to designate arm in weight and balance computation?

1—Pound/feet.
2—Inches.
3—Feet.
4—Pound/inches.

5178. What determines whether the value of the moment is preceded by a plus (+) or a minus (–) sign in aircraft weight and balance?

1—The addition or removal of weight.
2—The location of the weight in reference to the datum.
3—The result of a weight being added or removed and its location relative to the datum.
4—The location of the datum in reference to the aircraft CG.

5179. The maximum weight of an aircraft is the

1—empty weight plus crew, maximum fuel, cargo, and baggage.
2—empty weight plus crew, passengers, and fixed equipment.
3—empty weight plus useful load.
4—empty weight plus crew and fixed equipment.

5180. In the theory of weight and balance, the influence of weight is directly dependent upon its distance from the

1—centerline of the aircraft.
2—center of the fuselage.
3—CG.
4—center of longitudinal axis.

5181. When computing weight and balance for a helicopter, you must consider that

1—the flight altitudes of a helicopter are such that the weight and balance is not critical.
2—it is different from a fixed-wing aircraft, because the whirling rotor makes it difficult to locate the mean aerodynamic chord.
3—the arm of tail-mounted components is subject to constant change.
4—it is computed the same as a fixed-wing aircraft.

5182. Which of the following should be clearly indicated on the aircraft weighing form?

1—Minimum allowable gross weight.
2—Weight of unusable fuel.
3—Weighing points.
4—Zero fuel weight.

5183. The reference datum line is usually placed forward of the nose of the aircraft because

1—this is the only location from which all arms can be measured.
2—all manufacturers have agreed on this point for purposes of standardization.
3—all measurements will be positive numbers, contributing to accuracy.
4—measurement of arms from the nose involves less movement of cargo.

5184. Zero fuel weight is the

1—basic operating weight plus the weight of full passengers and cargo.
2—maximum weight of a loaded airplane less fuel located in the wings.
3—gross weight plus fuel, passengers, and cargo.
4—basic weight plus the weight of items such as oil, crew, and crew's baggage.

5185. The empty weight of an airplane is determined by

1—adding the gross weight on each weighing point and multiplying by the measured distance to the datum.
2—adding the net weight of each weighing point and multiplying the measured distance to the datum.
3—subtracting the tare weight from the scale reading and adding the weight of each weighing point.
4—multiplying the measured distance from each weighing point to the datum times the sum of scale reading less the tare weight.

5186. When dealing with weight and balance of an aircraft, the term maximum weight is interpreted to mean the maximum

1—weight of the empty aircraft.
2—weight of the useful load.
3—authorized weight of the aircraft and its contents.
4—weight of all optional or special equipment that can be installed in the aircraft.

5187. The most important reason for aircraft weight and balance control in today's aircraft is

1—efficiency in flight.
2—to reduce noise levels.
3—safety.
4—to increase payloads.

5188. An aircraft has an empty weight of 2,886 pounds and a total moment of 107,865.78 inch–pounds. The following modifications were made: two seats weighing 15 pounds each were removed from station 73 and replaced with a cabinet weighing 97 pounds; one seat weighing 20 pounds was installed at station 73; communication gear weighing 30 pounds was installed at station 97. The new empty weight CG is located

1—1.63 inches forward of the old CG.
2—2.05 inches forward of the old CG.
3—1.63 inches aft of the old CG.
4—2.05 inches aft of the old CG.

5189. When determining the empty weight of an aircraft, certificated under current airworthiness standards (FAR Part 23), the oil contained in the supply tank is considered

1—a part of the empty weight.
2—a part of the useful load.
3—the same as the fluid contained in the water injection reservoir.
4—a part of the payload.

5190. Improper loading of a helicopter which results in exceeding either the fore or aft CG limits is hazardous due to the

1—reduction or loss of effective cyclic pitch control.
2—override feature of the sprag clutch assembly.
3—Coriolis effect being translated to the fuselage.
4—reduction or loss of effective collective pitch control.

5191. How is the useful load of an aircraft determined?

1—Subtract the empty weight from the maximum allowable gross weight.
2—Add the empty weight and maximum allowable gross weight.
3—Subtract the tare weight from the empty weight.
4—Multiply the total moment by the empty weight.

5192. The maximum weight as used in weight and balance control of a given aircraft, can normally be found

1—in the back of the aircraft logbook.
2—by weighing the aircraft to obtain empty weight and mathematically adding the weight of fuel, oil, pilot, passengers, and baggage.
3—in the Aircraft Specifications or Type Certificate Data Sheets.
4—by adding the empty weight and payload.

5193. An aircraft with an empty weight of 2,100 pounds and an empty weight CG of +32.5 was altered as follows:

(1) two 18–pound passenger seats located at +73 were removed;
(2) structural modifications were made at +77 increasing weight by 17 pounds;
(3) a seat and safety belt weighing 25 pounds were installed at +74.5; and
(4) radio equipment weighing 35 pounds was installed at +95.

What is the new empty weight CG?

1—+30.44.
2—+34.01.
3—+33.68.
4—+34.65.

5194. The CG range in single rotor helicopters is

1—in a location that prevents external loads from being carried.
2—much greater than for airplanes.
3—approximately the same as the CG range for airplanes.
4—more restricted than for airplanes.

5195. The amount of fuel used for computing empty weight and corresponding CG is

1—empty fuel tanks.
2—unusable fuel.
3—full fuel tanks.
4—the amount of fuel necessary for 1/2 hour of operation.

5196. An aircraft as loaded weighs 4,954 pounds at a CG of +30.5 inches. The CG range is +32.0 inches to +42.1 inches. Find the minimum weight of the ballast necessary to bring the CG within the CG range. The ballast arm is +162 inches.

1—61.98 lb.
2—30.58 lb.
3—46.24 lb.
4—57.16 lb.

5197. As weighed, the total empty weight of an aircraft is 5,862 pounds with a moment of 885,957. However, when the aircraft was weighed, 20 pounds of alcohol were on board at +84, and 23 pounds of hydraulic fluid were in a tank located at +101. What is the empty weight CG of the aircraft?

1—150.700.
2—151.700.
3—154.200.
4—151.365.

5198. Two boxes which weigh 10 pounds and 5 pounds are placed in an airplane so that their distance aft from the CG are 4 feet and 2 feet respectively. How far forward of the CG should a third box, weighing 20 pounds, be placed so that the CG will not be changed?

1—3 feet.
2—2.5 feet.
3—6 feet.
4—8 feet.

5199. When the empty weight CG of an aircraft falls within the empty weight CG range listed in the Aircraft Specifications or Type Certificate Data Sheet, the

1—critical fore–and–aft–loaded CG positions computations are unnecessary.
2—aircraft must be loaded in accordance with nonstandard arrangements.
3—critical forward–loaded CG position must be computed.
4—critical rearward–loaded CG position must be computed.

5200. An aircraft with an empty weight of 1,800 pounds and an empty weight CG of +31.5 was altered as follows:

(1) two 15–pound passenger seats located at +72 were removed;
(2) structural modifications increasing the weight 14 pounds were made at +76;
(3) a seat and safety belt weighing 20 pounds were installed at +73.5; and
(4) radio equipment weighing 30 pounds was installed at +30.

What is the new empty weight CG?

1—+30.61.
2—+35.04.
3—+31.61.
4—+32.69.

5201. How is the moment of an item about the datum obtained?

1—Arm times the item weight.
2—Item weight times its distance from the loaded CG.
3—Item weight times its distance from the empty CG.
4—From the Aircraft Specifications or Type Certificate Data Sheets.

5202. If a 40–pound generator applies +1400 inch–pounds to a reference axis, the generator is located

1— –25 from the axis.
2— –35 from the axis.
3— +35 from the axis.
4— +25 from the axis.

5203. In a balance computation of an aircraft from which an item located aft of the datum was removed, use

1—(+)weight x (+)arm (+)moment.
2—(–)weight x (+)arm (–)moment.
3—(–)weight x (–)arm (+)moment.
4—(+)weight x (–)arm (–)moment.

5204.
Datum is forward of the
   main gear center point....................................30.24 in.
Actual distance between tail gear and
   main gear center points.................................360.26 in.
Net weight at right main gear..............................9,980 lb.
Net weight at left main gear...............................9,770 lb.
Net weight at tail gear...1,970 lb.

The following items were in the aircraft when weighed:

a. Lavatory water tank full (34 lb. at +352).
b. Hydraulic fluid (22 lb. at –8).
c. Removable ballast (146 lb. at +380).

What is the empty weight CG of the aircraft described above?

1—62.92 inches.
2—60.31 inches.
3—58.54 inches.
4—59.50 inches.

5205. When making a rearward weight and balance check to determine that the CG will not exceed the rearward limit during extreme conditions, the items of useful load which should be computed at their minimum weights are those located forward of the

1—forward CG limit.
2—empty weight CG.
3—datum.
4—rearward CG limit.

5206. When an empty aircraft is weighed, the combined net weight at the main gears is 3,540 pounds with an arm of 195.5 inches. At the nose gear, the net weight is 2,322 pounds with an arm of 83.5 inches. The datum line is forward of the nose of the aircraft. What is the empty CG of the aircraft?

1—151.1.
2—158.7.
3—155.2.
4—146.5.

5207. An aircraft with an empty weight of 1,500 pounds and an empty weight CG of +28.4 was altered as follows:

(1) two 12–pound seats located at +68.5 were removed;
(2) structural modifications weighing +28 pounds were made at +73;
(3) a seat and safety belt weighing 30 pounds were installed at +70.5; and
(4) radio equipment weighing 25 pounds was installed at +85.

What is the new empty weight CG?

1—+23.51.
2—+30.81.
3—+31.35.
4—+30.30.

5208. The following alteration was performed on an aircraft: A model B engine weighing 175 pounds was replaced by a model D engine weighing 185 pounds at a –62.00 inches station. The aircraft weight and balance records show the previous empty weight to be 998 pounds and an empty weight CG of 13.48 inches. What is the new empty weight CG?

1—12.99 inches.
2—13.96 inches.
3—14.25 inches.
4—12.73 inches.

5209. When computing the maximum forward loaded CG of an aircraft, minimum weights, arms, and moments should be used for items of useful load that are located aft of the

1—rearward CG limit.
2—forward CG limit.
3—datum.
4—empty weight CG.

5210. An aircraft with an empty weight of 1,800 pounds and an empty weight CG of +31.5 was altered as follows:

(1) two 15–pound passenger seats located at +50 were removed;
(2) structural modifications increasing the weight 14 pounds were made at +76;
(3) a seat and safety belt weighing 20 pounds were installed at +73.5; and
(4) radio equipment weighing 30 pounds was installed at +30.

What is the new empty weight CG?

1—+31.97.
2—+31.67.
3—+32.69.
4—+33.95.

5211. Find the empty weight CG location for the following tricycle gear aircraft. Each main wheel weighs 753 pounds, nosewheel weighs 22 pounds, distance between nosewheel and main wheels is 87.5 inches, nosewheel location is +9.875 inches from datum, with 1 gallon of hydraulic fluid at –21.0 inches included in the weight scale.

1—+97.375 inches.
2—+94.89 inches.
3—+95.61 inches.
4—+96.11 inches.

5212. Which coupling nut should be selected for use with 1/2–inch aluminum oil lines which are to be assembled using flared tube ends and standard AN nuts, sleeves, and fittings?

1—AN–818–2.
2—AN–818–8.
3—AN–818–5.
4—AN–818–12.

5213. Hydraulic lines located in entryways or passenger, crew, or baggage compartments

1—should be suitably supported and protected against physical damage.
2—are not normally permitted.
3—must be routed in separate enclosures which must be drained and vented to the outside atmosphere.
4—must not contain any fittings or connections within the entryways or compartments.

5214. From the following sequences of steps, indicate the proper order you would use to make a single flare on a piece of tubing:

A.Place the tube in the proper size hole in the flaring block.
B.Project the end of the tube slightly from the top of the flaring tool, about the thickness of a dime.
C.Slip the fitting nut and sleeve on the tube.
D.Strike the plunger several light blows with a light weight hammer or mallet and turn the plunger one–half turn after each blow.
E.Tighten the clamp bar securely to prevent slippage.
F.Center the plunger or flaring pin over the tube.

1—F, E, B, A, D, C.
2—A, C, E, B, D, F.
3—C, A, F, B, E, D.
4—C, B, F, E, A, D.

5215. High–pressure hydraulic tubing, which is damaged in a localized area to such extent that repair is necessary, may be repaired

1—with a union and two sets of connecting fittings.
2—only by replacing the entire tubing using the same size and material as the original.
3—by cutting out the damaged section of tubing and installing a short piece of high–pressure flexible hose with hose clamps.
4—by cutting out the damaged section and welding in a replacement section of tubing.

5216. What is an advantage of a double flare on aluminum tubing?

1—It is less concentric than a single flare.
2—Ease of construction.
3—It is less resistant to the shearing effect of torque.
4—It is more resistant to the shearing effect of torque.

5217. If failure of a flexible hydraulic hose equipped with swaged end fittings occurs, which of the following repair procedures should be followed?

1—Insert a nonmetallic hose liner which is approved for use with the type of fluid contained in the system and clamp firmly at both ends.
2—Remove the hose fittings and reuse on a new flexible line of the correct length.
3—Replace the hose with rigid tubing equipped with end fittings of the same type as those used in other parts of the system.
4—Install a replacement hose of the proper length which has been factory equipped with swaged end fittings.

5218. Select the correct statement in reference to flare fittings.

1—AN fittings can easily be identified by the shoulder between the end of the threads and the flare cone.
2—All parts of the AN fitting assemblies are interchangeable with AC fitting assemblies with the exception of the sleeves.
3—AC and AN fittings are identical except for the material from which they are made and the identifying color.
4—AC fittings have generally replaced the older AN fittings.

5219. Flexible lines should be installed

1—where bends are necessary.
2—only aft of the firewall.
3—with just enough slack to make the connection.
4—with 5 – 8 percent slack.

5220. The maximum distance between end fittings to which a straight hose assembly is to be connected is 50 inches. The minimum hose length to make such a connection should be

1—54–1/2 inches.
2—50–3/4 inches.
3—51 inches.
4—52–1/2 inches.

5221. How should a flexible hose be installed?

1—Stretched tightly between two fittings.
2—With slack or bend in the hose.
3—With as small a bend radius as possible.
4—To allow maximum flexing during operation.

5222. Which of the following colors identifies an AN aluminum flared–tube fitting?

1—Green.
2—Black.
3—Red.
4—Blue.

5223. Soft aluminum tubing (1100, 3003, or 5052) may be bent by hand if the size is

1—5/16 inch or less.
2—7/16 inch or less.
3—5/8 inch or less.
4—1/4 inch or less.

5224. The material specifications for a certain aircraft require that a replacement oil line be fabricated from 3/4–inch, 0.072 5052–0 aluminum alloy tubing. What is the inside dimension of this tubing?

1—0.606 inch.
2—0.572 inch.
3—0.688 inch.
4—0.750 inch.

5225. In most aircraft hydraulic systems, two–piece tube connectors consisting of a sleeve and a nut are used when a tubing flare is required. The use of this type connector eliminates

1—the need for flexible lines interconnecting movable and stationary components.
2—the flaring operation prior to assembly.
3—the possibility of reducing the flare thickness by wiping or ironing during the tightening process.
4—wrench damage to the tubing during the tightening process.

5226. Which of the following statements about MS (Military Standard) flareless fittings is correct?

1—MS flareless fitting sleeves must not be preset on the line prior to final assembly.
2—During installation, MS flareless fittings are normally tightened by turning the nut a specified amount after the sleeve and fitting sealing surface have made contact, rather than being torqued.
3—MS flareless fittings should not be lubricated prior to assembly.
4—MS flareless fittings must be tightened to a specific torque.

5227. When flaring aluminum tubing for use with AN coupling nuts and sleeves, the flare angle should be

1—30°.
2—37°.
3—67°.
4—45°.

5228. Scratches or nicks on the straight portion of aluminum alloy tubing may be repaired if they are no deeper than

1—20 percent of the wall thickness.
2—1/32 inch.
3—1/16 inch.
4—10 percent of the wall thickness.

5229. Which of the following is used in aircraft plumbing to connect moving parts with stationary parts in locations subject to vibration?

1—Flexible aluminum tubing.
2—Corrosion-resistant steel tubing.
3—Thin-wall aluminum tubing.
4—Flexible hose.

5230. If it is necessary to route hydraulic lines and electrical power cables adjacent to each other, the electrical cable should be

1—routed below the hydraulic line and clamped securely to the structure.
2—routed above the hydraulic line and clamped securely to the structure.
3—attached securely to the line by clamps that will maintain a separation between the line and cable equal to at least four times the diameter of the fluid line.
4—grounded at intervals not to exceed four times the distance separating the fluid line and the cable.

5231. When installing a hydraulic line, the tube flare should meet the fitting squarely and fully before starting the nut. The flare should never be drawn to the fitting with the nut because this may

1—deform the line.
2—strip the threads on the fitting.
3—deform the flare.
4—strip the threads on the nut.

5232. Flexible hose used in aircraft plumbing is classified in size according to the

1—outside diameter.
2—cross-sectional area.
3—wall thickness.
4—inside diameter.

5233. A scratch or nick in aluminum alloy tubing can be repaired by burnishing provided the scratch or nick does not

1—exceed 5 percent of the tube diameter.
2—appear in the heel of a bend in the tube.
3—exceed 20 percent of the wall thickness of the tube.
4—exceed 10 percent of the tube diameter.

5234. What is the advantage of using Teflon hose in aircraft fluid systems?

1—Lower operating temperatures.
2—Its operating strength.
3—Nonresistant to fatigue.
4—Easier to inspect.

5235. Which of the following tubings has the characteristics (medium strength, workability) necessary for use in a medium-pressure (1,500 PSI) hydraulic system?

1—2024T or 5052-0 aluminum alloy.
2—Copper or hard plastic.
3—1100-1/2H or 3003-1/2H aluminum alloy.
4—Carbon steel or chrome molybdenum.

5236. Which of the following tubings has the characteristics (high strength, abrasion resistance) necessary for use in a high-pressure (3,000 PSI) hydraulic system for operation of landing gear and flaps?

1—Copper or hard plastic.
2—2024-T or 5052-0 aluminum alloy.
3—Corrosion-resistant steel annealed or 1/4H.
4—1100-1/2H or 3003-1/2H aluminum alloy.

5237. When installing bonded clamps to support metal tubing,

1—paint removal from tube is unnecessary as it will inhibit corrosion.
2—paint clamp and tube after clamp installation for slippage identification.
3—paint clamp and tube after clamp installation to prevent corrosion.
4—remove paint or anodizing from tube at clamp location.

5238. In a metal tubing installation,

1—rigid straight line runs are preferable.
2—tension is undesirable because pressurization will cause it to expand and shift.
3—a tube may be pulled in line if nut will start on threaded coupling.
4—tension is desirable because it cannot shift when pressurized.

5239. A fluid line marked with the letters PHDAN

1—carries hydraulic fluid back to the reservoir.
2—is a high–pressure fluid line. The letters mean Pressure High, Discharge at Nacelle.
3—is carrying a fluid which may be dangerous to personnel.
4—must be made of a nonphosphorous metal.

5240. The principal advantage of Teflon hose is

1—its operating strength.
2—its low cost.
3—its corrosion resistance.
4—its flexibility.

5241. Which of the following statements concerning Bernoulli's principle is true?

1—The pressure of a fluid increases at points where the velocity of the fluid increases.
2—The pressure of a fluid decreases at points where the velocity of the fluid increases.
3—It has no practical application in today's aircraft.
4—It applies only to gases.

5242. (1) Bonded clamps are used for support when installing metal tubing.

(2) Unbonded clamps are used for support when installing wiring.

Concerning the above statements,

1—only No. 2 is true.
2—only No. 1 is true.
3—both No. 1 and No. 2 are true.
4—neither No. 1 nor No. 2 is true.

5243. Flexible hose may be used in aircraft fluid systems

1—to replace only low–pressure fluid system lines.
2—to replace any fluid system line not subject to heat.
3—according to the manufacturer's specifications.
4—to replace any fluid system line.

5244. Which of the following nondestructive inspection methods would be most suitable for inspecting wing internal structures?

1—Continuous magnetic particle inspection.
2—Penetrant inspection.
3—Radiographic inspection.
4—Residual magnetic particle inspection.

5245. Magnetic particle inspection is used primarily to detect

1—distortion.
2—irregular surfaces.
3—porosity and thickness of ferromagnetic metals.
4—flaws on or near the surface.

5246. In order for dye penetrant inspection to be effective, the material being checked must

1—have subsurface cracks.
2—be magnetic.
3—be nonmagnetic.
4—have surface cracks.

5247. Which of the following nondestructive testing methods is most successful in detecting intergranular corrosion in nonferrous metal?

1—Spectography inspection.
2—Radiography inspection.
3—Magnetic particle inspection.
4—Eddy current inspection.

5248. Which of the following nondestructive inspection methods would normally be the most satisfactory to determine the internal structural condition of a highly stressed cast aluminum alloy fitting?

1—X–ray or radiographic inspection.
2—Dye penetrant inspection.
3—Magnetic particle inspection.
4—Fluorescent penetrant inspection.

5249. (1) Magnetic particle inspection is a method of detecting invisible cracks and other defects in ferrous metals.

(2) Penetrant inspection is a nondestructive test for defects open to the surface in parts made of nonporous material.

Regarding the above statements, which one of the following is true?

1—Only No. 1 is true.
2—Only No. 2 is true.
3—Both No. 1 and No. 2 are true.
4—Neither No. 1 nor No. 2 is true.

5250. What is the principal advantage of the radiographic inspection method of nondestructive testing?

1—Minimum safety precautions required.
2—Little or no disassembly of structure.
3—Simplicity of equipment operation.
4—Low cost.

5251. What method of magnetic particle inspection is used most often to inspect aircraft parts for invisible cracks and other defects?

1—Residual.
2—Inductance.
3—Continuous.
4—Intermittent.

5252. Which of the following factors is considered basic knowledge of X–ray exposure?

1—The density, thickness, size, and shape of the object to be exposed.
2—Processing of the film.
3—The distance and angle for 1–minute exposure.
4—The type and exact size of the discontinuity to be detected.

5253. Which method of magnetization produces the greatest sensitivity in locating subsurface discontinuities?

1—Continuous circular with a solenoid.
2—Continuous longitudinal with a solenoid.
3—Residual circular with a conductor bar.
4—Residual longitudinal with a conductor bar.

5254. The magnetic particle inspection is somewhat unreliable for distinguishing or detecting

1—laps and fatigue cracks.
2—cold shuts.
3—hairline discontinuities and small cracks.
4—discontinuities which are below the surface.

5255. What type of discontinuities are formed by a parting or a rupture of the metal being magnetically tested?

1—Lap or voids.
2—Cold shut and inclusions.
3—Inclusion and laps.
4—Pipes or voids.

5256. Which statement relating to the residual magnetizing inspection method is true?

1—It requires careful and intelligent interpretation and evaluation of the discontinuities it reveals.
2—It is used in practically all circular and longitudinal magnetizing procedures.
3—It may be used only with steels which have been heat treated for stressed applications.
4—Its indicating medium is applied when the magnetizing force is being maintained.

5257. What two types of indicating mediums are available for magnetic particle inspection?

1—Iron and ferric oxides.
2—Wet and dry process materials.
3—Penetrant and fluorescent material.
4—High retentivity and low permeability material.

5258. Which of the following metals can be inspected using the magnetic particle inspection method?

1—Magnesium alloys.
2—Aluminum alloys.
3—Iron alloys.
4—Copper.

5259. How is an item demagnetized after a magnetic particle inspection?

1—Rinse it with a nonmagnetic solution.
2—Lay it aside for a period of time.
3—Subject it to an alternating current energized coil.
4—Subject the item to direct current.

5260. Which type crack can be detected by magnetic particle inspection using either circular or longitudinal magnetization?

1—45°.
2—Longitudinal.
3—Transverse.
4—Circumferential.

5261. Surface cracks in aluminum castings and forgings may usually be detected by

1—the use of dye penetrants and suitable developers.
2—heating the part to approximately 750° and observing the surface for any material that may have been forced out of a crack.
3—magnetic particle inspection.
4—submerging the part in a concentrated solution of caustic soda (sodium hydroxide) and rinsing with clear water.

5262. To detect a minute crack using dye penetrant inspection requires

1—that the developer be applied to a wet surface.
2—a shorter than normal penetrating time.
3—a longer than normal penetrating time.
4—the surface to be highly polished.

5263. When checking an item with the magnetic particle inspection method, circular and longitudinal magnetization should be used to

1—reveal all possible defects.
2—prevent permanent magnetization.
3—prevent one–way polarization.
4—ensure uniform current flow.

5264. What is the primary limitation of the dye penetrant method of inspection?

1—The defect must be open to the surface.
2—The smaller the defect, the longer the penetrating time required.
3—The washing or rinsing operation causes unreliable results.
4—It is limited in use to a small number of metals.

5265. What is the purpose of the developer as used in dye penetrant inspection?

1—It removes penetrant from surface.
2—It causes penetrant to glow in the dark.
3—It causes penetrant on surface to seep into cracks.
4—It brings out the penetrant to show defects.

5266. Dye penetrant inspection method will detect

1—surface defects.
2—subsurface flaws.
3—deteriorated molded rubber.
4—cracks in any porous material.

5267. Before applying dye penetrant to an aluminum part, first

1—clean the part thoroughly.
2—apply the developer.
3—mix the penetrant with water.
4—mix the penetrant with the developer.

5268. If dye penetrant inspection indications are not sharp and clear, the most probable cause is that

1—the part was not correctly degaussed before the developer was applied.
2—the part is not damaged.
3—the part is badly damaged over an extensive area of its surface.
4—the part was not thoroughly washed before developer was applied.

5269. Which type of current is usually used to demagnetize aircraft parts?

1—Alternating, decreasing.
2—Direct, increasing.
3—Alternating, increasing.
4—Direct, decreasing.

5270. When using the dye penetrant test, it is well to remember that the

1—smaller the defect, the longer the required penetrating time.
2—larger the defect, the longer the required penetrating time.
3—smaller the defect, the shorter the required penetrating time.
4—penetrating time is independent of the size of the defect.

5271. If a pure metal is heated above its critical temperature and cooled to room temperature, it will

1—form a mechanical mixture.
2—return to its original structure.
3—form a combination of a solid solution and mechanical mixture.
4—form a complex solution.

5272. The pattern for an inclusion is a magnetic particle buildup forming

1—a fernlike pattern.
2—a single line.
3—a smooth outline.
4—parallel lines.

5273. A part which is being prepared for dye penetrant inspection should be cleaned

1—by sandblasting.
2—with a volatile petroleum–base solvent.
3—with the penetrant developer.
4—with water–base solvents only.

5274. Which type crack will probably cause the most buildup in the magnetic particle indicating medium?

1—Heat–treated.
2—Shrink.
3—Fatigue.
4—Grinding.

5275. Under magnetic particle inspection, a part will be identified as having a fatigue crack under which of the following conditions?

1—The discontinuity pattern is straight.
2—The discontinuity is found in a non–stressed area of the part.
3—The discontinuity is found in a highly stressed area of the part.
4—The discontinuity pattern is not clear.

5276. The main disadvantage of dye penetrant inspection is that

1—the chemicals used are dangerous to the inspection personnel.
2—the defect must be open to the surface.
3—it does not work on nonferrous metals.
4—it is excessively time consuming.

5277. What defects will be detected by magnetizing a part using continuous longitudinal magnetization with a cable?

1—Defects perpendicular to the long axis of the part.
2—Defects parallel to the long axis of the part.
3—Defects perpendicular to the concentric circles of magnetic force within the part.
4—Defects parallel to the concentric circles of magnetic force within the part.

5278. Circular magnetization of a part can be used to detect which of the following defects?

1—Defects parallel to the long axis of the part.
2—Defects perpendicular to the long axis of the part.
3—Defects perpendicular to the concentric circles of magnetic force within the part.
4—Defects parallel to the concentric circles of magnetic force within the part.

5279. What type of corrosion attacks grain boundaries of aluminum alloys which are improperly or inadequately heat–treated?

1—Stress.
2—Intergranular.
3—Surface.
4—Fretting.

5280. If too much time is allowed to elapse during the transfer of 2017 or 2024 aluminum alloy from the heat treatment medium to the quench tank, it may result in

1—case hardening.
2—a dull, stained, or streaked surface.
3—retarded age hardening.
4—impaired corrosion resistance.

5281. Which heat–treating process of metal produces a hard, wear–resistant surface over a strong, tough core?

1—Case hardening.
2—Annealing.
3—Tempering.
4—Normalizing.

5282. Which heat–treating operation would be performed when the surface of the metal is changed chemically by introducing a high carbide or nitride content?

1—Tempering.
2—Normalizing.
3—Casehardening.
4—Annealing.

5283. Transfer of 2024 aluminum alloy from the heat–treat medium to the quench tank must be done quickly to

1—prevent stress cracking.
2—retard the age–hardening process.
3—attain good corrosion–resistant qualities.
4—ensure a good bond between core and aluminum coating.

5284. When heat treating a small aluminum forging, a cold water quench should be used to produce

1—maximum hardness.
2—minimum distortion.
3—minimum diffusion.
4—maximum corrosion resistance.

5285. Nitriding is a process which

1—forms a hard case on a part to resist wear.
2—decreases the size of the grain structure.
3—toughens steel to increase its tensile strength.
4—increases bearing heat resistance.

5286. In order to successfully heat treat ferrous metals, the rate of cooling is controlled by

1—allowing a time lag between soaking and quenching.
2—selecting a suitable quenching media.
3—artificial aging.
4—re–precipitation.

5287. The difference between annealing and normalizing steel is that normalizing usually requires temperatures

1—higher than critical point, with slower cooling.
2—lower than critical point, with faster cooling.
3—lower than critical point, with slower cooling.
4—higher than critical point, with faster cooling.

5288. Why is steel tempered after being hardened?

1—To increase its hardness and ductility.
2—To decrease its ductility and brittleness.
3—To increase its strength and decrease its internal stresses.
4—To relieve its internal stresses and reduce its brittleness.

5289. Hot or boiling water is normally used for quenching

1—tubing.
2—extrusions.
3—clad sheet.
4—large forgings.

5290. Which of the following aluminum alloy designations indicates that the metal has been solution heat treated and artificially aged?

1—3003–F.
2—7075–T6.
3—5052–H36.
4—6061–0.

5291. Alloy 2024 rivets should be driven within 10 minutes after they have been quenched; otherwise, before being used, they must be

1—aged.
2—normalized.
3—reheat treated.
4—refrigerated.

5292. Which of the following qualities of magnesium alloy are improved by precipitation heat treatment?

1—Ductility and toughness.
2—Ductility and yield strength.
3—Hardness and yield strength.
4—Hardness, ductility, and yield strength.

5293. The length of time that a piece of steel should be kept at the tempering temperature is determined by

1—type of steel.
2—thickness of the piece.
3—desired tensile strength.
4—quenching medium used in hardening.

5294. Which type of steel is best suited for case hardening?

1—Low–carbon, low–alloy.
2—High–carbon, low–alloy.
3—Low–carbon, high–alloy.
4—High–carbon, high–alloy.

5295. Which of the following cannot be heat treated repeatedly without harmful effects?

1—Unclad aluminum alloy in sheet form.
2—Products which are molded of steel.
3—6061–T9 stainless steel.
4—Clad aluminum alloy.

5296. Metal to be tested on the Rockwell Hardness Tester must meet which of the following specification(s)?

1—It must be ground smooth on one side only and cleaned.
2—The two opposite ground surfaces must be free of scratches and foreign matter.
3—The two opposite surfaces must be parallel but not necessarily cleaned.
4—The two opposite ground surfaces should be perpendicular and cleaned.

5297. What will determine the hardness of steel at ordinary temperatures?

1—The number of particles of iron carbide in the mixture.
2—The size of the particles of iron austenite in the mixture.
3—The distribution of the particles of iron matrix throughout the mixture.
4—The transformation of pearlite to austenite in the mixture.

5298. Unless otherwise specified, torque values for tightening aircraft nuts and bolts relate to

1—dry, thoroughly degreased threads.
2—lightly oiled threads.
3—well oiled threads.
4—threads lubricated with a dry lubricant.

5299. Which of the following is generally used in the construction of aircraft engine firewalls?

1—Aluminum alloy sheet.
2—Stainless steel.
3—Chrome–molybdenum alloy steel.
4—Magnesium–titanium alloy steel.

5300. Unless otherwise specified or required, aircraft bolts should be installed so that the bolt head is

1—upward, or in a rearward direction.
2—upward, or in a forward direction.
3—downward, or in a forward direction.
4—downward, or in a rearward direction.

5301. Alclad is a metal consisting of

1—aluminum alloy surface layers and a pure aluminum core.
2—pure aluminum surface layers on an aluminum alloy core.
3—a homogeneous mixture of pure aluminum and aluminum alloy.
4—alternating layers of pure aluminum and aluminum alloy.

5302. A fiber–type, self–locking nut should never be used on an aircraft if the bolt is

1—under shear loading.
2—under tension loading.
3—subject to rotation.
4—to be mounted in a vertical position.

5303. The Society of Automotive Engineers and the American Iron and Steel Institute use a numerical index system to identify the composition of various steels. The symbol 1020 indicates a plain carbon steel containing an average of

1—20.00 percent carbon.
2—2.00 percent carbon by volume.
3—2.00 percent carbon by weight.
4—0.20 percent carbon by weight.

5304. Which of the bolt head code markings shown in Figure 35 identifies a corrosion resistant AN standard steel bolt?

1—A.
2—B.
3—C.
4—D.

Figure 35

5305. Which of the following statements regarding aircraft bolts is correct?

1—AN standard steel bolts are marked with two raised dashes on the bolt head.
2—When tightening castellated nuts on drilled bolts, if the cotter pin holes do not line up, it is permissible to overtighten the nut to permit alignment of the next slot with the cotter pin hole.
3—In general, bolt grip lengths should equal the material thickness.
4—Alloy steel bolts smaller than 1/4-inch diameter should not be used in primary structure.

5306. Generally speaking, bolt grip lengths should be

1—one and one–half times the thickness of the material through which they extend.
2—equal to the thickness of the material through which they extend plus approximately one diameter.
3—equal to the thickness of the material through which they extend.
4—at least three times the thickness of the thinnest sheet.

5307. Select the grip length of a bolt used to fasten two 5/8–inch steel plates together.

1—1–3/8 inches.
2—1–1/4 inches.
3—3/4 inch.
4—7/8 inch.

5308. When the specific torque value for nuts is not given, where can the recommended torque value be found?

1—Advisory Circular 43.13–2A.
2—FAR Part 43, appendix D.
3—Technical Standard Order.
4—Advisory Circular 43.13–1A.

5309. Identify the clevis bolt illustrated in Figure 36.

1—A.
2—B.
3—D.
4—C.

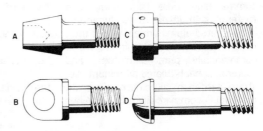

Figure 36

5310. A bolt with a single raised dash on the head is classified as an

1—AN corrosion–resistant steel bolt.
2—NAS standard aircraft bolt.
3—NAS close tolerance bolt.
4—AN aluminum bolt.

5311. How is a clevis bolt used with a fork–end cable terminal secured?

1—With a self–locking nut with a lock washer to prevent rotation of the bolt in the fork.
2—With a shear nut tightened to a snug fit, but with no strain imposed on the fork and safetied with a cotter pin.
3—With a castle nut tightened until slight binding occurs between the fork and the fitting to which it is being attached.
4—With a shear nut and cotter pin or a thin self–locking nut tightened enough to prevent rotation of the bolt in the fork.

5312. A cross inside a triangle on the head of a bolt used on approved aircraft installations would indicate the bolt is a

1—standard corrosion–resistant bolt.
2—special purpose bolt.
3—magnetically inspected bolt.
4—close tolerance bolt.

5313. Where is an AN clevis bolt used in an airplane?

1—In landing gear assemblies.
2—For tension and shear load conditions.
3—Where external tension loads are applied.
4—Only for shear load applications.

5314. A bolt with an X inside a triangle on the head is classified as an

1—AN aluminum alloy bolt.
2—NAS standard aircraft bolt.
3—NAS close tolerance bolt.
4—AN corrosion–resistant steel bolt.

5315. The core material of ALCLAD 2024–T4 is

1—commercially pure aluminum, and the surface material is strain–hardened aluminum alloy.
2—heat–treated aluminum alloy, and the surface material is commercially pure aluminum.
3—commercially pure aluminum, and the surface material is heat–treated aluminum alloy.
4—strain–hardened aluminum alloy, and the surface material is commercially pure aluminum.

5316. The high purity aluminum code number 1100 identifies what type of aluminum?

1—Aluminum alloy containing 11 percent copper.
2—Heat–treated aluminum alloy.
3—Aluminum alloy containing zinc.
4—99 percent commercially pure aluminum.

5317. Aircraft bolts are usually manufactured with a

1—class 1 fit for the threads.
2—class 2 fit for the threads.
3—class 3 fit for the threads.
4—class 4 fit for the threads.

5318. In the four–digit, aluminum index system number 2024, the first digit indicates

1—zinc has been added to the aluminum.
2—the percent of alloy added.
3—the different alloys in that group.
4—copper is the major alloying element.

5319. How is the locking feature of the fiber–type lock nut obtained?

1—By a saw–cut fiber insert with a pinched–in thread in the locking section.
2—By the use of an unthreaded fiber locking insert.
3—By a fiber insert held firmly in place at the base of the load carrying section.
4—By placing the threads in the fiber insert out–of–phase with the load carrying section.

5320. Identify the weld in Figure 37, caused by an excessive amount of acetylene.

1—D.
2—A.
3—C.
4—B.

5321. Using Figure 37, select the illustration which depicts a cold weld.

1—C.
2—A.
3—B.
4—D.

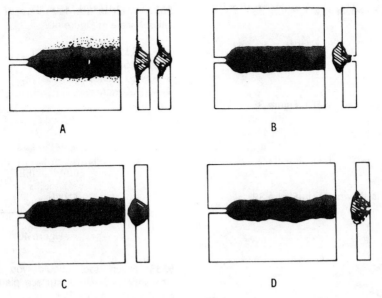

A   B

C   D

Figure 37

5322. If it is necessary to reweld a joint,

1—use an oversize welding rod.
2—remove all the old weld.
3—preheat before rewelding.
4—increase the temperature to ensure proper penetration.

5323. Why is it considered good practice to normalize a part after welding?

1—To relieve internal stresses developed within the base metal.
2—To burn out any excess carbon deposited during welding.
3—To introduce a slight amount of carbon to improve the surface hardness of the weld.
4—To remove the surface scale formed during welding.

5324. Holes and a few projecting globules are found in a weld. What action should be taken?

1—Fill the holes with solder.
2—Reheat the bead to melt the globules.
3—Remove all the old weld and reweld the joint.
4—File the rough surface.

5325. Select a characteristic of a good gas weld.

1—The depth of penetration shall be sufficient to ensure fusion of the filler rod.
2—The weld should be built up 1/8 inch.
3—The weld should taper off smoothly into the base metal.
4—No oxide should be formed on the base metal.

5326. Which of the following is an unacceptable characteristic of a weld on an aircraft?

1—Projecting globules.
2—Smooth seam.
3—A smooth taper into the base metal.
4—An extra thickness built up at the joint.

5327. What is the most important characteristic that contributes to a sound weld?

1—Inclusions.
2—Fusion.
3—Puddling.
4—Porosity.

5328. One characteristic of a good weld is that no oxide should be formed on the base metal at a distance from the weld of more than

1—1/2 inch.
2—1 inch.
3—1/8 inch.
4—1/4 inch.

5329. What type weld is shown at A in Figure 38?

1—Fillet.
2—Butt.
3—Lap.
4—Edge.

5330. What type weld is shown at B in Figure 38?

1—Fillet.
2—Double butt.
3—Flat.
4—Edge.

5331. What type weld is shown at G in Figure 38?

1—Lap.
2—Butt.
3—Flat.
4—Edge.

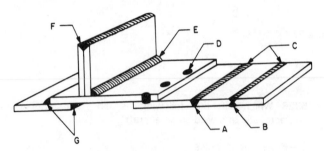

Figure 38

5332. Which tool can be used to measure axle end play?

1—Depth gauge.
2—Dial indicator.
3—Axle gauge.
4—Protractor.

5333. The measurement reading on the illustrated micrometer in Figure 39 is

1—0.2915.
2—0.2861.
3—0.2911.
4—0.2901.

Figure 39

5334. What is the measurement reading on the vernier caliper scale in Figure 40?

1—1.411 inches.
2—1.436 inches.
3—1.461 inches.
4—1.700 inches.

Figure 40

5335. Which tool should you use to measure the clearance between a surface plate and a surface being checked for flatness?

1—Depth gauge.
2—Surface gauge.
3—Thickness gauge.
4—Dial indicator.

5336. If the thimble of a standard micrometer caliper, graduated in thousandths of an inch, is turned one full revolution, the spindle will move

1—0.010 inch.
2—0.040 inch.
3—1.000 inch.
4—0.025 inch.

5337. Which of the following represents the vernier scale graduation of a micrometer?

1—.00001.
2—.001.
3—.0001.
4—.01.

5338. What is the reading of the micrometer illustrated in B of Figure 41?

1—.2252.
2—.2152.
3—.2552.
4—.2557.

5339. Using Figure 41, select the illustration of the micrometer that reads .2403.

1—A.
2—C.
3—B.
4—D.

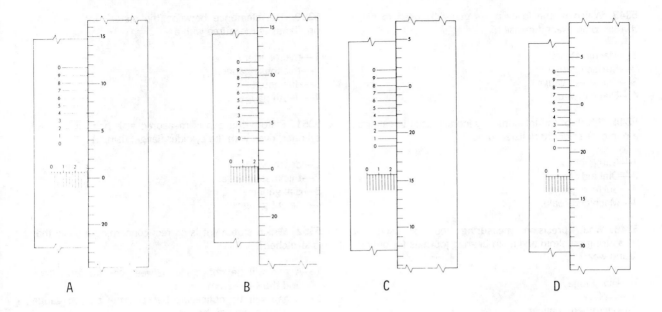

A   B   C   D

Figure 41

5340. Which tool is used to find the center of a shaft or other cylindrical work?

1—Combination set.
2—Surface gauge.
3—Dial indicator.
4—Micrometer caliper.

5341. If it is necessary to accurately measure the diameter of a hole approximately 3/8 inch in diameter, the mechanic should use a

1—telescoping gauge and read the measurement directly from the gauge.
2—telescoping gauge and determine the size of the hole by taking a micrometer reading of the adjustable end of the telescoping gauge.
3—0– to 1–inch inside micrometer and read the measurement directly from the micrometer.
4—small–hole gauge and determine the size of the hole by taking a micrometer reading of the ball end of the gauge.

5342. The measurement reading on the micrometer appearing in Figure 42 is

1—.2958.
2—.2902.
3—.2812.
4—.2885.

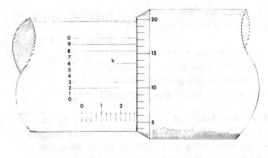

Figure 42

5343. Which of the following is generally used to set a divider to an exact dimension?

1—Machinist scale.
2—Surface gauge.
3—Thickness gauge.
4—Dial indicator.

5344. Which of the following is generally used to calibrate a micrometer or check its accuracy?

1—Gauge block.
2—Dial indicator.
3—Surface gauge.
4—Machinist scale.

5345. What precision measuring tool is used for measuring crankpin and main bearing journals for out–of–round wear?

1—Dial gauge.
2—V–blocks.
3—Micrometer caliper.
4—Depth gauge.

5346. The side clearance of piston rings is measured with a

1—depth gauge.
2—thickness gauge.
3—hole gauge.
4—telescopic gauge.

5347. Crankshaft alignment runout can be checked by rotating the shaft on V–blocks and a surface plate. The measurements are taken with a

1—height gauge.
2—dial gauge.
3—depth gauge.
4—telescopic gauge.

5348. How can the dimensional inspection of a bearing in a rocker arm be accomplished?

1—Depth gauge and micrometer.
2—Thickness gauge and V–blocks.
3—Thickness gauge and push–fit arbor.
4—Telescopic gauge and micrometer.

5349. The twist of a connecting rod is checked by installing push–fit arbors in both ends, supported by parallel steel bars on a surface plate. Measurements are taken between the arbor and the parallel bar with a

1—dial gauge.
2—height gauge.
3—thickness gauge.
4—depth gauge.

5350. The clearance between the piston rings and the ring lands is measured with a

1—micrometer caliper.
2—thickness gauge.
3—dial gauge.
4—depth gauge.

5351. Piston–ring gap is measured at a point inside the cylinder, even with the cylinder flange, using a

1—depth gauge.
2—thickness gauge.
3—dial gauge.
4—height gauge.

5352. Which statement is correct concerning a valve that is stretched?

1—A gap will be noticeable between the stretch gauge and the valve stem.
2—A gap will be noticeable between the stretch gauge and the radius of the valve.
3—A gap will be noticeable between the stretch gauge and both the valve stem and the valve face.
4—The stretch gauge will touch only in the radius that connects the valve stem to the valve face.

5353. Which tool can be used to determine piston pin out–of–round wear?

1—Telescopic gauge.
2—Micrometer caliper.
3—Dividers.
4—Dial indicator.

5354. During starting of a turbojet powerplant using a compressed air starter, a hung start occurred. Select the proper procedure from the following.

1—Advance the power lever.
2—Increase airpower to the starter.
3—Re–engage the starter.
4—Shut the engine down.

5355. When towing an aircraft,

1—discharge all hydraulic pressure to prevent accidental operation of the landing gear retracting mechanism.
2—all tailwheel aircraft must be towed backwards.
3—if the aircraft has a steerable nosewheel, the locking scissors should be set to full swivel.
4—all nosewheel aircraft must be towed backwards.

5356. When starting an aircraft engine equipped with a float–type carburetor, the carburetor air heat control should be placed in

1—the COLD position.
2—the COLD position under non–icing conditions and in the HOT position under icing conditions.
3—the HOT position to reduce the engine warmup period.
4—a position between HOT and COLD.

5357. When approaching the front of an idling turbojet engine, the hazard area extends forward of the engine approximately

1—35 feet.
2—5 feet.
3—15 feet.
4—25 feet.

5358. The most satisfactory extinguishing agent for use in case of carburetor or intake fire is

1—dry chemical.
2—carbon tetrachloride.
3—carbon dioxide.
4—fine water spray.

5359. Identify the signal to engage rotor on a rotorcraft in Figure 43.

1—A.
2—B.
3—D.
4—C.

A   B

C   D

Figure 43

5360. Starting a reciprocating engine which has a liquid lock in one of the cylinders could cause

1—compressive failure of the valve bodies.
2—starter clutch damage.
3—removal of the lubricating film on the cylinder walls.
4—major structural damage.

5361. The priming of a fuel injected horizontally opposed engine is accomplished by placing the fuel control lever in which of the following?

1—AUTO–LEAN position.
2—IDLE–CUTOFF position.
3—AUTO–RICH position.
4—FULL–RICH position.

5362. Which of the following can be a cause of preignition?

1—High intake manifold pressure.
2—Vaporized fuel compressed to a critical pressure.
3—Feathered edges on valves.
4—Excessive spark plug gap.

5363. Which action would be required if the turbine inlet temperature exceeds the specified maximum during the starting sequence of a turboprop engine?

1—Turn off the fuel and ignition switch, discontinue the start and make an investigation.
2—Advance the power lever and observe for excessive smoke, if present, discontinue the start.
3—Continue the start since temperature will stabilize as soon as 5,000 RPM is reached.
4—Turn off the fuel and ignition switch, discontinue the start, then wait 5 minutes and initiate the starting sequence.

5364. How is a flooded engine, equipped with a float–type carburetor, cleared of excessive fuel?

1—Crank the engine with the starter or by hand, with the mixture control in cutoff, ignition switch off, and the throttle fully open, until the fuel charge has been cleared.
2—Crank the engine with the starter or by hand, with the mixture control in cutoff, ignition switch off, and the throttle closed, until the fuel has been cleared.
3—Turn off the fuel and the ignition. Discontinue the starting attempt until the excess fuel has cleared.
4—Crank the engine with the starter or by hand, with the mixture control in cutoff, ignition switch on, and the throttle fully open, until the excess fuel has cleared or until the engine starts.

5365. If fuel flows steadily from the internal supercharger drain valve during an attempt to start an engine, which of the following is a possible cause of the trouble?

1—Excessive booster pump pressure.
2—The fuel pump relief valve is out of adjustment.
3—An improper setting of the mixture control.
4—A fatigued carburetor fuel supply line.

5366. Which of the hand signals in Figure 44, would you give if a taxiing aircraft were in danger of striking some object?

1—C.
2—A.
3—B.
4—D.

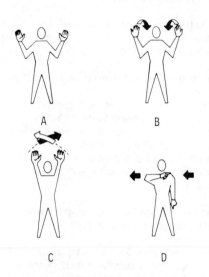

A   B

C   D

Figure 44

5367. Induction fires during starting can be extinguished by

1—directing carbon dioxide into the air intake of the engine.
2—directing carbon dioxide into the exhaust system.
3—closing the fuel shutoff valve.
4—closing the throttle.

5368. Which of the following indicators should be checked immediately after starting a reciprocating engine?

1—Oil pressure.
2—Manifold pressure.
3—Tachometer.
4—Cylinder head temperature.

5369. During the starting sequence of a turboprop engine, there is no oil pressure indication at 5,000 RPM for either the reduction gear or the power unit. What action is required?

1—Turn off the ignition switch and fuel, discontinue the start, and replace the oil pressure gauge.
2—Turn off the ignition switch and fuel, let the engine motorize for 5 minutes, and restart.
3—Turn off the ignition switch and fuel, discontinue the start, and make an investigation.
4—Turn off the ignition switch and fuel, discontinue the start, wait for 5 minutes, and initiate the starting sequence.

5370. When starting an engine equipped with a float–type carburetor with an idle cutoff unit, the mechanic should place the mixture control in the

1—IDLE–CUTOFF position.
2—FULL–LEAN position.
3—FULL–RICH position while priming the engine; however, the mixture control should be returned to the IDLE–CUTOFF position when actually starting the engine.
4—FULL–RICH position.

5371. When approaching the rear of an idling turbojet engine, the hazard area extends aft of the engine approximately

1—200 feet.
2—100 feet.
3—25 feet.
4—5 feet.

5372. During starting of a turbojet powerplant using a compressed air starter, a hot start occurrence was recorded. Select what happened from the following.

1—The pneumatic starting unit overheated.
2—The pneumatic starter overheated.
3—The powerplant was preheated before starting.
4—The fuel/air mixture was excessively rich.

5373. Aviation gasoline mixed with jet fuel will have the following effects on turbine powerplant efficiency.

1—No disadvantages.
2—The tetraethyl lead in the gasoline forms deposits on the turbine blades.
3—The tetraethyl lead in the gasoline forms deposits on the compressor blades.
4—Continuous use will not affect engine efficiency.

5374. What would be the action to take if aviation gasoline in an aircraft was contaminated with jet fuel?

1—Determine the type jet fuel, then contact the engine manufacturer for guidance.
2—Adjust mixture to compensate for lower octane.
3—Determine the amount of fuel put in tanks and if under 25 percent of total volume, do nothing. If over 25 percent, drain the tanks and refill.
4—Drain fuel tanks and fill with proper octane rated fuel.

5375. Jet fuel is more susceptible to contamination than aviation gasoline.

(1) Jet fuel is of higher viscosity and therefore holds contaminants better.

(2) Viscosity has no relation to contamination of fuel.

Regarding the above statements, which of the following is true?

1—Only No. 1 is true.
2—Only No. 2 is true.
3—Both No. 1 and No. 2 are true.
4—Neither No. 1 nor No. 2 is true.

5376. Identify the color of low–lead 100LL aviation gasoline.

1—Blue.
2—Green.
3—Orange.
4—Purple.

5377. What is the identification color for turbine fuel?

1—Blue.
2—Green.
3—Straw.
4—Purple.

5378. How are aviation fuels, which possess greater antiknock qualities than 100 octane, classified?

1—As antidetonants.
2—According to the milliliters of lead.
3—By reference to normal heptane.
4—By performance numbers.

5379. Why is ethylene dibromide added to all grades of aviation gasoline?

1—To prevent detonation at takeoff power settings.
2—To remove zinc silicate deposits from the spark plugs.
3—To scavenge lead oxide from the cylinder combustion chambers.
4—To increase the antiknock rating of the fuel.

5380. Both gasoline and kerosene have certain advantages for use as turbine fuel. Which of the following is a true statement in reference to the advantages of each?

1—Kerosene has a higher heat energy per unit weight than gasoline.
2—Gasoline is a better lubricant than kerosene (important in regard to fuel metering pumps).
3—Gasoline has a higher heat energy per unit volume than kerosene.
4—Kerosene has a higher heat energy per unit volume than gasoline.

5381. What characteristic of a fuel reduces its tendency to vapor lock?

1—Low latent heat of vaporization.
2—High volatility.
3—Low vapor pressure.
4—High octane rating or performance number.

5382. Characteristics of detonation are:

1—rapid rise in cylinder pressure, excessive cylinder head temperature, and an increase in engine power.
2—cylinder pressure remains the same, excessive cylinder head temperature, and a decrease in engine power.
3—rapid rise in cylinder pressure, excessive cylinder head temperature, and a decrease in engine power.
4—rapid rise in cylinder pressure, cylinder head temperature normal, and a decrease in engine power.

5383. A fuel that vaporizes too readily may cause

1—hard starting.
2—detonation.
3—slow warmup.
4—vapor lock.

5384. Jet fuel number identifiers are

1—performance numbers to designate the volatility of the fuel.
2—performance numbers and are relative to the fuel's performance in the aircraft engine.
3—type numbers and have no relation to the fuel's performance in the aircraft engine.
4—type numbers used to designate the vaporization rate at atmospheric pressure.

5385. The second number in a fuel rating (100/130) is the

1—lean mixture antiknock rating.
2—percent of iso–octane in the fuel.
3—percent of normal heptane in the fuel.
4—rich mixture antiknock rating.

5386. Characteristics of aviation gasoline are

1—high heat value, high volatility.
2—high heat value, low volatility.
3—low heat value, high volatility.
4—low heat value, low volatility.

5387. Tetraethyl lead is added to gasoline to

1—retard the formation of corrosives.
2—lubricate the valve seats.
3—improve the gasoline's performance in the engine.
4—dissolve the moisture in the gasoline.

5388. A fuel that does not vaporize readily enough can cause

1—vapor lock.
2—detonation.
3—hard starting.
4—surface ignition.

5389. Which of the following materials would be used to clean magnesium engine parts prior to painting?

1—Dichromate solution.
2—Acetone.
3—MEK (methyl ethyl ketone).
4—20 percent caustic soda solution.

5390. How may magnesium engine parts be cleaned?

1—Soak in a 20 percent caustic soda solution.
2—Wash with gasoline.
3—Spray with MEK (methyl ethyl ketone).
4—Wash with a commercial solvent, decarbonize, and scrape or grit blast.

5391. Which solvent is recommended for removing grease from fabric prior to doping?

1—Kerosene.
2—Acetone.
3—Toluene.
4—Turpentine.

5392. When an anodized surface coating is damaged in service, it can be partially restored by

1—use of a metal polish.
2—chemical surface treatment.
3—complete penetration of an inhibitor.
4—a suitable mild cleaner.

5393. If a nickel–cadmium battery electrolyte is spilled on the hands, what agent may be used to rinse the affected area?

1—Boric acid solution.
2—Baking soda solution.
3—Ammonia solution.
4—Potassium hydroxide solution.

5394. Select the solvent recommended for wipedown of cleaned surfaces just before painting.

1—Aliphatic naphtha.
2—Trichloroethane.
3—Dry–cleaning solvent.
4—Kerosene.

5395. Nickel–cadmium battery cases and drain surfaces which have been affected by electrolyte should be neutralized with a solution of

1—boric acid.
2—diluted sulphuric acid.
3—sodium bicarbonate.
4—potassium hydroxide.

5396. Which of the following is used for general cleaning of aluminum surfaces by mechanical means?

1—Carborundum paper.
2—Aluminum wool.
3—Crocus cloth.
4—Steel wool.

5397. Select the solvent used to clean acrylics and rubber.

1—Aliphatic naphtha.
2—Methyl ethyl ketone.
3—Aromatic naphtha.
4—Methyl chloroform.

5398. Select the solvent used to clean metal surfaces and as a paint stripper for small areas.

1—Kerosene.
2—Methyl ethyl ketone.
3—Dry–cleaning solvent.
4—Methyl chloroform.

5399. Fayed surfaces cause concern in chemical cleaning because of the danger of

1—electrostatic charge buildup.
2—electrochemical attack.
3—entrapping corrosive materials.
4—corrosion by imbedded iron oxide.

5400. What effect does a caustic cleaning product have on aluminum structures?

1—Electrochemical.
2—Mechanical.
3—Strengthening.
4—Corrosive.

5401. Fretting corrosion between closely fitting metal parts

1—is cause for rejection of the part.
2—may occur when steel bushings are pressed into aluminum parts with too tight a fit.
3—occurs only when two dissimilar metals are in contact.
4—may be repaired by light oiling and retightening of the parts.

5402. The lifting or flaking of a metal surface is known as exfoliation. This delamination of the grain boundaries is sometimes known as

1—dissimilar metal corrosion.
2—intergranular corrosion.
3—stress corrosion.
4—climatic condition corrosion.

5403. Select the metal on which corrosion forms a greenish film.

1—Copper and its alloy.
2—Aluminum and its alloy.
3—Steel and iron.
4—Titanium and its alloy.

5404. Alodizing is a chemical treatment for aluminum alloys to improve their paint–bonding qualities and to

1—decrease their corrosion resistance.
2—make their surface slightly alkaline.
3—relieve their surface stresses.
4—increase their corrosion resistance.

5405. Metals and alloys of greatly differing composition should not be in direct contact with each other because

1—of the possibility of generating static charges which interfere with radio reception.
2—of the different rates of expansion.
3—of their unequal tensile strengths.
4—deterioration may result from electrochemical action at point of contact.

5406. What is the best means of removing rust from steel?

1—Mechanical.
2—Electrical.
3—Chemical.
4—Biochemical.

5407. How is alodine applied to aluminum alloys?

1—As part of the priming process.
2—By dipping.
3—As part of the manufacturing process.
4—Concurrently with the alclad.

5408. The lifting or flaking of the metal at the surface due to delamination of grain boundaries caused by the pressure of corrosion residual product buildup is called

1—brinelling.
2—electrolysis.
3—transgranulation.
4—exfoliation.

5409. The artificial production of a film of hydroxide on the surface of aluminum or any of its alloys is commonly called

1—alodizing.
2—parco lubrizing.
3—anodizing.
4—dichromating.

5410. Why are parts rinsed thoroughly in hot water after they have been heat–treated in a sodium and potassium nitrate bath?

1—To prevent corrosion.
2—To prevent blistering.
3—To reduce warpage.
4—To retard discoloration.

5411. Intergranular corrosion in structural aluminum alloy parts

1—is not likely to occur in parts fabricated from heat–treated sheet aluminum.
2—may be detected by the white, powdery deposit formed on the surface of the metal.
3—is not likely to occur in parts fabricated from aluminum–coated alloys (ALCLAD or PURECLAD).
4—cannot always be detected by surface indications.

5412. Intergranular corrosion is caused by

1—improper heat treatment.
2—dissimilar metal contact.
3—improperly assembled components.
4—poor application of zinc chromate primer.

5413. Which of the following types of corrosion may exist in aircraft structure and not be visible?

1—Electrolytic corrosion.
2—Bonderizing corrosion.
3—Climatic condition corrosion.
4—Surface corrosion.

5414. Corrosion should be removed from magnesium parts with a

1—steel wire brush.
2—carborundum abrasive.
3—stiff, hog–bristle brush.
4—steel burnishing tool.

5415. Why is it important not to rotate the propeller shaft after the final spraying of corrosion–preventive mixture into cylinders installed on removed engines?

1—The link rods may be damaged by hydraulic lock.
2—The corrosion preventive mixture will be excessively diluted.
3—The engine may fire and cause injury to personnel.
4—The seal of corrosion preventive mixture will be broken.

5416. Why is a plastic surface flushed with fresh water before it is cleaned with soap and water?

1—To prevent crazing.
2—To prevent scratching.
3—To remove oil and grease.
4—To prevent softening the plastic.

5417. What should be done to prevent rapid deterioration when a tire becomes covered with lubricating oil?

1—Wipe the tire with a dry cloth, and then dry with compressed air.
2—Wipe the tire with a dry cloth followed by a washdown with soap and water.
3—Wash the tire with a petroleum solvent, and then dry with compressed air.
4—Wash the tire with alcohol or lacquer thinner to neutralize the action of the oil.

5418. Galvanic action caused by dissimilar metal contact may best be prevented by

1—applying a nonporous dielectric material between the surfaces.
2—priming both surfaces with a light coat of zinc chromate primer.
3—cleaning both surfaces with MEK (methyl ethyl ketone).
4—application of paper tape between the surfaces.

5419. Corrosion caused by electrolytic action is the result of

1—excessive anodization.
2—contact between two unlike metals.
3—the wrong quenching agent.
4—excessive etching.

5420. To prevent corrosion between dissimilar metal joints in which magnesium alloy is involved,

1—prime only the magnesium part with one coat of zinc chromate primer and put a leather gasket between the pieces.
2—the magnesium part must be plated with the same metal as the part it will be in contact with.
3—coat both mating surfaces with an aluminized varnish.
4—prime both parts with two coats of zinc chromate primer and place a layer of pressure sensitive vinyl tape between them.

5421. The interior surface of sealed structural steel tubing is best protected against corrosion by

1—purging the interior with dry air before sealing.
2—a coating of hot linseed oil.
3—evacuating the tubing before sealing.
4—a coating of zinc chromate paint.

5422. Failure to quench a piece of aluminum in the minimum required time could result in

1—failure of the ALCLAD to adhere properly.
2—causing the aluminum to become extremely brittle.
3—impairing the corrosion resisting qualities of the metal.
4—difficulty in preparing the surface for painting.

5423. What power of ten is equal to 1,000,000?

1—10 to the third power.
2—10 to the fourth power.
3—10 to the fifth power.
4—10 to the sixth power.

5424. Find the square root of 1,746.

1—41.7852.
2—41.7752.
3—40.7742.
4—42.7854.

5425. The result of nine raised to the fourth power is which of the following?

1—6,491.
2—6,461.
3—6,941.
4—6,561.

5426 Find the square root of 3,722.1835.

1—61.00971.
2—61.00.
3—60.009.
4—61.0097.

5427. Find the square of 212.

1—40,144.
2—43,924.
3—44,944.
4—44,844.

5428. Find the value of ten raised to the negative sixth power.

1—0.000001.
2—0.000010.
3—0.0001.
4—0.00001.

5429. What is the square root of four raised to the fifth power?

1—32.
2—64.
3—16.
4—20.

5430. The number 3.47 X 10 to the negative fourth power is equal to

1—3,470.00.
2—34,700.0.
3—0034.70.
4—.000347.

5431. Which alternative answer is equal to 16,300?

1—1.63 X 10 to the fourth power.
2—1.63 X 10 to the negative third power.
3—163 X 10 to the negative second power.
4—1,630 X 10 to the negative first power.

5432. Find the square root of 125.131.

1—111.8 X 10 to the third power.
2—.1118 X 10 to the negative second power.
3—1,118 X 10 to the negative second power.
4—11.186 X 10 to the negative third power.

5433. What is the square root of 16 raised to the fourth power?

1—1,024.
2—64.
3—4,096.
4—256.

5434. Solve the following:

$$\frac{\sqrt[2]{31} + \sqrt[2]{43}}{(17)^2}$$

1—.0297.
2—.1680.
3—.7132.
4—.0419.

5435. The result of seven raised to the third power plus the square root of 39 is equal to

1—349.24.
2—27.24.
3—343.24.
4—2,407.24.

5436. Find the square root of 1,824.

1—42.708 X 10 to the negative second power.
2—.42708.
3—4,270.8.
4—.42708 X 10 to the second power.

5437. The total piston displacement of a specific engine is

1—dependent on the compression ratio.
2—dependent on the volumetric efficiency.
3—the volume displaced by all the pistons during one revolution of the crankshaft.
4—the volume displaced by one piston during one revolution of the crankshaft.

5438. Compute the area of the trapezoid in Figure 45.

1—40 square feet.
2—32 square feet.
3—240 square feet.
4—960 square feet.

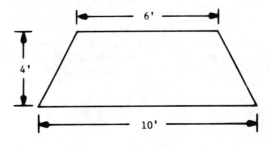

Figure 45

5439. What size sheet of metal is required to fabricate a cylinder 20 inches long and 8 inches in diameter?

Note: C = 3.1416 X d

1—20″ X 25–5/32″.
2—20″ X 24–9/64″.
3—20″ X 25–9/64″.
4—20″ X 24–5/32″.

5440. Find the area of the right triangle shown in Figure 46.

1—5 sq. in.
2—6 sq. in.
3—9 sq. in.
4—12 sq. in.

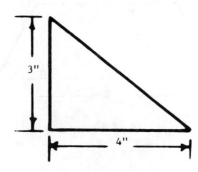

Figure 46

5441. What force is exerted on the piston in a hydraulic cylinder if the area of the piston is 1.2 square inches and the fluid pressure is 850 PSI?

1—1,020 lb.
2—1,220 lb.
3—960 lb.
4—850 lb.

5442. The cylinder of an aircraft engine has a bore of 4.75 inches and a stroke of 4.5 inches. What is the piston displacement of one cylinder?

1—64.34 cu. in.
2—33.58 cu. in.
3—318.81 cu. in.
4—79.74 cu. in.

5443. A rectangular–shaped fuel tank measures 60 inches in length, 30 inches in width, and 12 inches in depth. How many cubic feet are within the tank?

1—12.5.
2—15.0.
3—18.5.
4—21.0.

5444. Select the container size that will be equal in volume to 60 gallons of fuel. (7.5 gal. = 1 cu. ft.)

1—7.0 cu. ft.
2—7.5 cu. ft.
3—8.0 cu. ft.
4—8.5 cu. ft.

5445. Compute the area of the trapezoid in Figure 47.

1—8 sq. ft.
2—24 sq. ft.
3—48 sq. ft.
4—10 sq. ft.

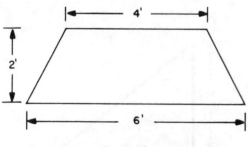

Figure 47

5446. Determine the area of the triangle formed by points A, B, and C in Figure 48.

A to B = 7.5 inches
A to D = 16.8 inches

1—7.07 sq. in.
2—42 sq. in.
3—63 sq. in.
4—126 sq. in.

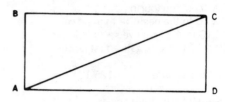

Figure 48

5447. What is the piston displacement of a master cylinder with a 1.5–inch diameter bore and a piston stroke of 4 inches?

1—1.7671 cu. in.
2—9.4247 cu. in.
3—7.0685 cu. in.
4—6.1541 cu. in.

5448. How many gallons of fuel will be contained in a rectangular–shaped tank which measures 2 feet in width, 3 feet in length, and 1 foot 8 inches in depth?
(7.5 gal. = 1 cu. ft.)

1—66.6.
2—75.
3—110.
4—45.

5449. A rectangular–shaped fuel tank measures 27–1/2 inches in length, 9 inches in width, and 8–1/4 inches in depth. How many gallons will the tank contain?
(231 cu. in. = 1 gal.)

1—18.8.
2—6.4.
3—8.8.
4—16.4.

5450. A four–cylinder aircraft engine has a cylinder bore of 3.78 inches and is 8.5 inches deep. With the piston on bottom center, the top of the piston measures 4.0 inches from the bottom of the cylinder. What is the approximate piston displacement of this engine?

1—200 cu. in.
2—360 cu. in.
3—320 cu. in.
4—235 cu. in.

5451. A rectangular–shaped fuel tank measures 37–1/2 inches in length, 14 inches in width, and 8–1/4 inches in depth. How many cubic inches are within the tank?

1—525.
2—433.125.
3—4,331.25.
4—309.375.

5452. Select the fraction which is equal to .020.

1—3/16.
2—1/5.
3—2/7.
4—1/50.

5453. Select the decimal which is equal to the mixed number 1–7/32.

1—1.2188.
2—1.3932.
3—1.7320.
4—1.3270.

5454. If the volume of a cylinder with the piston at bottom center is 84 cubic inches and the piston displacement is 70 cubic inches, then the compression ratio is

1—7:1.
2—1.2:1.
3—6:1.
4—1.9:1.

5455. Express 7/8 as a percent.

1—.785 percent.
2—78.5 percent.
3—.875 percent.
4—87.5 percent.

5456. What is the speed of a spur gear with 42 teeth driven by a pinion gear with 14 teeth turning 420 RPM?

1—14 RPM.
2—42 RPM.
3—160 RPM.
4—140 RPM.

5457. An engine develops 108 horsepower at 87 percent power. What horsepower would be developed at 65 percent power?

1—80.
2—94.
3—70.
4—64.

5458. Which of the following alternatives is the decimal equivalent of the fraction 43/32?

1—1.34375.
2—1.43325.
3—1.32435.
4—1.74415.

5459. Select the fractional equivalent for a 0.09375–thick sheet of aluminum.

1—5/64.
2—5/32.
3—3/64.
4—3/32.

5460. Express 5/8 as a percent.

1—62 percent.
2—.625 percent.
3—.620 percent.
4—62.5 percent.

5461. Select the decimal which is most nearly equal to 77/64.

1—1.8311.
2—0.8311.
3—1.2031.
4—1.3120.

5462. An airplane flying a distance of 875 miles used 70 gallons of gasoline. How many gallons will it need to travel 3,000 miles?

1—108 gallons.
2—120 gallons.
3—240 gallons.
4—144 gallons.

5463. What is the speed ratio of a gear with 36 teeth meshed to a gear with 20 teeth?

1—12 to 5.
2—12 to 6.6.
3—9 to 5.
4—17 to 10.

5464. A pinion gear with 14 teeth is driving a spur gear with 42 teeth at 140 RPM. Determine the speed of the pinion gear.

1—588 RPM.
2—196 RPM.
3—420 RPM.
4—240 RPM.

5465. The parts department's profit is 12 percent on a new magneto. How much does the magneto cost if the selling price is $145.60?

1—$128.12.
2—$120.00.
3—$125.60.
4—$130.00.

5466. An engine of 125 hp. maximum is running at 65 percent power. What is the hp. being developed?

1—93.05.
2—30.85.
3—81.25.
4—38.85.

5467. An engine of 98 hp. maximum is running at 75 percent power. What is the hp. being developed?

1—87.00.
2—33.30.
3—73.50.
4—41.30.

5468. A blueprint shows a hole of 0.17187 to be drilled. Which fraction size drill bit is most nearly equal?

1—11/64.
2—9/64.
3—9/32.
4—11/32.

5469. Which of the following decimals is most nearly equal to a bend radius of 31/64?

1—0.0645.
2—0.6450.
3—0.48437.
4—0.3164.

5470. Sixty–five engines are what percent of 80 engines?

1—71 percent.
2—81 percent.
3—65 percent.
4—52 percent.

5471. The radius of a piece of round stock is 7/32. Select the decimal which is most nearly equal to the diameter.

1—0.2187.
2—0.5343.
3—0.4375.
4—0.3531.

5472. Maximum engine life is 900 hours. Recently, 27 engines were removed with an average life of 635.30 hours. What percent of the maximum engine life has been achieved?

1—71 percent.
2—72 percent.
3—73 percent.
4—74 percent.

5473. What is the ratio of 10 feet to 30 inches?

1—4:1.
2—1:3.
3—1:4.
4—3:1.

5474. How much current does a 30–volt motor, 1/2 hp., 85 percent efficient, draw from the bus?

(Note: 1 hp. = 746 watts.)

1—14.6 amperes.
2—12.4 amperes.
3—12.3 amperes.
4—14.1 amperes.

5475. Solve the following:

$$[(4 \text{ X} -3) + (-9 \text{ X} +2)] \div 2 =$$

1— 0.
2— −30.
3— −15.
4— −5.

5476. Solve the following:

$$(64 \text{ X } 3/8) \div 3/4 =$$

1—16.
2—24.
3—32.
4—72.

5477. Solve the following:

$$-18.4 + 0.07 -2.2 + 8.36 =$$

1— +12.17.
2— −12.31.
3— +12.31.
4— −12.17.

5478. Solve the following:

$$1/2 + 7/8 -1/3 =$$

1—1–11/16.
2—1–1/24.
3—1–1/12.
4—1–3/8.

5479. Solve the following:

$(1{-}1/2 \times 7/8) + 1/4 =$

1—1–11/16.
2—1–3/4.
3—1–9/16.
4—1–17/32.

5480. Solve the following:

$(32 \times 3/8) \div 1/6 =$

1—16.
2—12.
3—2.
4—72.

5481. What is the ratio of a gasoline fuel load of 200 gallons to one of 1,680 pounds?

1—3:7.
2—5:7.
3—2:3.
4—5:42.

5482. Solve the following:

$(2{-}1/2 \times 1{-}3/8) + 2{-}3/4 =$

1—6–3/16.
2—6–1/4.
3—3–5/16.
4—4–3/4.

5483. Solve the following:

$5/9 \, (41 + 40) - 40 =$

1—5.
2—4.55.
3—22.3.
4—73.8.

5484. Solve the following:

$(32 \times 3/16) \div 1/2 =$

1—16.
2—12.
3—72.
4—1.5.

5485. Solve the following:

$2/4 \, (30 + 34) \, 5 =$

1—117.
2—160.
3—345.
4—640.

5486. Solve the equation in Figure 49.

1— 16.51.
2— 174.85.
3— –81.49.
4— 14.00.

$$\frac{(-35 + 25)\,(-7) + (\pi)\,(16^{-2})}{\sqrt{25}} =$$

Figure 49

5487. Solve the following equation:

$$-4\,\overline{\smash{\big)}\,125}$$

$$-6\,\overline{\smash{\big)}\,{-36}}$$

1— +31.25.
2— –5.20.
3— –31.25.
4— +6.05.

5488. Solve the following:

$4 - 3\left[-6(2+3) + 4\right] =$

1—82.
2—33.
3— –25.
4— –71.

5489. Solve the following:

$-6\left[-9(-8+4) - 2(7+3)\right] =$

1— 16.
2— –332.
3— 216.
4— –96.

5490. Solve the following:

$5[4 - (+3-4) + 13] - 6 =$

1—84.
2—21.
3—17.
4— –16.

5491. Which of the following provides a place for indicating compliance with FAA Airworthiness Directives or manufacturers' service bulletins?

1—Structural repair manual.
2—Aircraft overhaul manual.
3—Aircraft maintenance records.
4—Illustrated parts catalog.

5492. An aircraft was not approved for return to service after an annual inspection and the owner wanted to fly the aircraft to another maintenance base. Which statement is correct?

1—The owner must obtain a special flight permit.
2—The aircraft must be repaired and approved prior to any flight.
3—The aircraft may be flown to another maintenance base if the discrepancies are not "safety of flight" items.
4—The owner must obtain a restricted category type certificate.

5493. When should special inspection procedures be followed to determine if any damage to the aircraft structure has occurred?

1—Progressive inspection.
2—100-hour inspection.
3—Overweight landing.
4—Annual inspection.

5494. During an annual inspection, if a defect is found which makes the aircraft unairworthy, the person disapproving must

1—remove the airworthiness certificate from the aircraft.
2—submit a Malfunction or Defect Report.
3—provide a written notice of the defect to the owner.
4—repair the defect before completion of the inspection.

5495. What is the means by which the FAA notifies aircraft owners and other interested persons of unsafe conditions and prescribes the condition under which the product may continue to be operated?

1—Airworthiness Directives.
2—Supplemental Type Certificates.
3—Malfunction or Defect Reports.
4—Technical Standard Orders.

5496. Where should you find this entry?

"Removed right wing from aircraft and removed skin from outer 6 feet. Repaired buckled spar 49 inches from tip in accordance with Figure 8 in the manufacturer's structural repair manual No. 28–1."

1—Aircraft engine maintenance record.
2—AD compliance record.
3—FAA Form 337.
4—Service bulletin compliance record.

5497. Which of the following is a powerplant major repair?

1—Installation of an accessory which is not approved for the engine.
2—Separation of a crankcase of a reciprocating engine equipped with an integral supercharger.
3—Overhaul of pressure-type carburetors, and pressure-type fuel and oil pumps.
4—Disassembly of a crankshaft of a reciprocating engine equipped with a spur-type propeller reduction gear.

5498. Which of the following is an airframe major repair?

1—Removal, installation, and repair of landing gear tires.
2—Changes to the wing or to fixed or movable control surfaces which affect flutter and vibration characteristics.
3—Rewinding the field coil of an electrical accessory.
4—The repair of portions of skin sheets by making additional seams.

5499. What maintenance action should follow if the maintenance record contained the following entry?

"During flight the aircraft engine experienced sudden deceleration."

1—Perform power output check (static and idle RPM).
2—Perform the engine group inspection for the annual/progressive.
3—Perform power output check with vibration analyzer (static and idle RPM).
4—Perform sudden stoppage inspection.

5500. Which of the following aircraft record entries best describes a repair of a 6–1/4–inch crack in the top skin of a stabilizer?

1—Stop-drilled the crack. Make permanent repair at next scheduled annual inspection.
2—Stop-drilled the crack using a 1/8-inch drill, installed a patch plate two times the length of the crack, two-row rivet spacing two times the diameter of the rivet with a rivet edge distance of three times the diameter of the rivet used.
3—Cut out the defective area, installed a flush repair using the same rivet size, spacing materials, and pattern as the adjacent area.
4—Cut out the defective area, installed a flush repair using the same material, rivet size, and unit spacing as the adjacent area and a rivet edge distance of two times the diameter of the rivets used.

5501. Which of the following statements best describes the entry made in an aircraft record for the repair of a 2–inch crack located in the center of a plastic side window?

1—Stop–drilled both ends of the crack, cut a piece of plastic 1–1/2 inches larger than the crack, heated the patch in an oil bath and formed to the contour to be patched, applied plastic adhesive, pressed in place, and let dry under pressure for 3 hours.
2—Stop–drilled both ends of the crack, cut a piece of plastic 3/4 inch larger than the crack, heated the patch in an oil bath and formed to the contour to be patched, applied plastic adhesive, pressed in place, and let dry under pressure for 3 hours.
3—Stop–drilled both ends of the crack, cut a piece of plastic 3/4 inch larger than the crack, heated the patch in an oil bath or boiling water and formed to the contour to be patched, and applied plastic adhesive, pressed in place, and let dry under pressure for 3 hours.
4—Stop–drilled both ends of the crack. Glued a piece of plastic over the crack.

5502. Which of the following aircraft record entries best describes a repair of a dent in a tubular steel structure dented at a cluster?

1—Removed and replaced all the members of the cluster.
2—Removed and replaced the damaged member.
3—Welded a reinforcing plate over the dented area.
4—Filled the damaged area with a molten metal and dressed to the original contour.

5503. Who is responsible for making the entry in the maintenance records after an annual, 100–hour, or progressive inspection?

1—Any certificated airframe mechanic.
2—The owner or operator of the aircraft.
3—The person approving or disapproving for return to service.
4—The pilot performing the test flight.

5504. An aircraft owner was provided a list of discrepancies on an aircraft that was not approved for return to service after an annual inspection. Which statement is correct concerning who may correct the discrepancies?

1—Only a mechanic with an Inspection Authorization.
2—An appropriately rated mechanic.
3—Any certificated repair station.
4—The owner or operator of the aircraft.

5505. When entering a major repair in an aircraft logbook, a mechanic must enter

1—the date the major repair was started.
2—total time in service for the aircraft.
3—his or her name and date the work was performed.
4—total aircraft time since the last overhaul.

5506. What action is required when a minor repair is performed on a certificated aircraft?

1—An FAA Form 337 must be completed.
2—The aircraft specifications must be amended.
3—An entry in the aircraft's permanent records is required.
4—The owner of the aircraft must report the repair to the FAA.

5507. After making a major repair to an aircraft engine that is to be returned to service, FAA Form 337, Major Repair and Alteration, must be prepared. How many copies are required and what is the disposition of the completed forms?

1—Two; both copies for the FAA.
2—Two; one copy for the aircraft owner and one copy for the FAA.
3—Three; one copy for the aircraft owner and two copies for the FAA.
4—Three; one copy for the aircraft owner, one copy for the FAA, and one copy for the permanent records of the repairing agency or individual.

5508. Who is responsible for maintaining the required maintenance records for an airplane?

1—Authorized inspector.
2—Repair station operator.
3—Certificated mechanic.
4—Aircraft owner.

5509. Each person performing an annual or 100–hour inspection shall use a checklist that contains at least those items in the appendix of

1—FAR Part 43.
2—FAR Part 65.
3—FAR Part 91.
4—Advisory Circular 43.13–3.

5510. An FAA Form 337 is used to record and document which of the following?

1—Preventive and routine maintenance.
2—Major repairs and major alterations.
3—Minor repairs and minor alterations.
4—Airworthiness Directive compliance.

5511. After a mechanic holding an airframe and powerplant rating completes a 100–hour inspection, what action is required before the aircraft is returned to service?

1—Obtain a renewal for the airworthiness certificate.
2—Make the proper entries in the aircraft's maintenance record.
3—Complete an operational check of all systems.
4—A mechanic with an Inspection Authorization must approve the inspection.

5512. Which statement is correct concerning the completion of FAA Form 337?

1—The certificate number of the person approving the major repair or alteration need not be recorded on the form.
2—Weight and balance computations are not necessary if the empty weight change is less than 1 percent.
3—A clear, concise, and legible description of the work to be accomplished.
4—Flight test results should be recorded on the form.

5513. Each person performing an annual or 100–hour inspection shall use a checklist that contains at least those items in FAR Part 43, appendix

1—D.
2—A.
3—B.
4—C.

5514. A certificated airframe and powerplant mechanic is authorized to approve for return to service an aircraft after a

1—100–hour inspection.
2—certification inspection.
3—major alteration.
4—major repair.

5515. The force that can be produced by an actuating cylinder whose piston has a cross–sectional area of 3 square inches operating in a 1,000 PSI hydraulic system is most nearly

1—300 pounds.
2—3,000 pounds.
3—334 pounds.
4—1,000 pounds.

5516. The boiling point of a given liquid varies

1—inversely with volume.
2—directly with pressure.
3—inversely with pressure.
4—directly with volume.

5517. One of the following equals 1 horsepower.

1—2,000 ft./lb. of work per minute.
2—550 ft./lb. of work per minute.
3—2,000 ft./lb. of work per second.
4—33,000 ft./lb. of work per minute.

5518. Which of the following is not considered a method of heat transfer?

1—Convection.
2—Conduction.
3—Diffusion.
4—Radiation.

5519. An engine that weighs 350 pounds is removed from an aircraft by means of a mobile hoist. The engine is raised 3 feet above its attachment mount, and the entire assembly is then moved forward 12 feet. A constant force of 70 pounds is required to move the loaded hoist. What is the total work input required to move the hoist?

1—840 ft./lb.
2—22,680 ft./lb.
3—1,890 ft./lb.
4—1,050 ft./lb.

5520. Which of the following is the actual amount of water vapor in a mixture of air and water?

1—Relative humidity.
2—Dew point.
3—Absolute humidity.
4—Vapor pressure.

5521. Under which of the following conditions will the rate of flow of a liquid through a metering orifice (or jet) be the greatest?

1—Unmetered pressure – 18 PSI,
   metered pressure – 17.5 PSI,
   atmospheric pressure – 14.5 PSI.
2—Unmetered pressure – 23 PSI,
   metered pressure – 12 PSI,
   atmospheric pressure – 14.3 PSI.
3—Unmetered pressure – 17 PSI,
   metered pressure – 5 PSI,
   atmospheric pressure – 14.7 PSI.
4—Unmetered pressure – 15 PSI,
   metered pressure – 12 PSI,
   atmospheric pressure – 14.7 PSI.

5522. Which of the following is the portion of atmospheric pressure exerted by the moisture in the air?

1—Absolute humidity.
2—Relative humidity.
3—Dew point.
4—Vapor pressure.

5523. Using Figure 50, the amount of force applied to rope A to lift the weight is

1—12 pounds.
2—15 pounds.
3—20 pounds.
4—30 pounds.

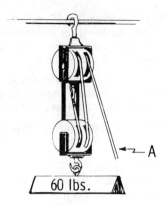

Figure 50

5524. Which of the following will weigh the least?

1—98 parts of dry air and 2 parts of water vapor.
2—35 parts of water vapor and 65 parts of dry air.
3—100 parts of dry air.
4—50 parts of dry air and 50 parts of water vapor.

5525. Which of the following is the ratio of the water vapor actually present in the atmosphere to the amount that would be present if the air were saturated at the prevailing temperature and pressure?

1—Absolute humidity.
2—Relative humidity.
3—Dew point.
4—Vapor pressure.

5526. If a 120–pound force moves a 3,200 pound object 20 feet in 20 seconds along a straight horizontal path, how much power was expended?

1—100 ft./lb. per second.
2—2,400 ft./lb. per second.
3—3,200 ft./lb. per second.
4—120 ft./lb. per second.

5527. If the container volume of a confined gas is doubled (assume temperature remains constant), the pressure will

1—increase in direct proportion to the volume increase.
2—remain the same.
3—be doubled.
4—be reduced to one–half its original value.

5528. If the temperature of a confined liquid is held constant and its pressure is tripled, the volume will

1—triple.
2—be increased one–third its original volume.
3—be reduced to one–third its original volume.
4—remain the same.

5529. If the pressure of an air mass is doubled, with all other conditions constant, its density

1—is doubled.
2—varies inversely as the pressure increases.
3—is decreased.
4—remains the same.

5530. How much work input is required to lower (not drop) a 120–pound weight from the top of a 3–foot table to the floor?

1—120 lb. of force.
2—120 ft./lb.
3—360 ft./lb.
4—None.

5531. Which of the following atmospheric conditions will cause the true landing speed of an aircraft to be the greatest?

1—Low temperature with high humidity.
2—Low temperature with low humidity.
3—High temperature with low humidity.
4—High temperature with high humidity.

5532. If the fluid pressure is 800 PSI in a 1/2–inch line supplying an actuating cylinder with a piston area of 10 square inches, the force exerted on the piston will be

1—4,000 pounds.
2—1,600 pounds.
3—8,000 pounds.
4—800 pounds.

5533. How many of the following factors are necessary to determine power?

(1) Force exerted.
(2) Distance the force moves.
(3) Time required to do the work.

1—None.
2—One.
3—Two.
4—Three.

5534. What force must be applied to roll a 120–pound barrel up an inclined plane 9 feet long to a height of 3 feet (disregard friction)?

$L \div I = R \div E$
L = Length of ramp, measured along the slope.
I = Height of ramp.
R = Weight of object to be raised or lowered.
E = Force required to raise or lower object.

1—40 lb.
2—120 lb.
3—360 lb.
4—393 lb.

5535. Which of the following statements concerning the concepts of heat and temperature is true?

1—There is a direct relationship between temperature and heat.
2—There is an inverse variation between temperature and heat.
3—Temperature is a measure of intensity of heat.
4—Temperature is a measure of quality of heat.

5536. What is absolute humidity?

1—The portion of atmospheric pressure that is exerted by the moisture in the air.
2—The temperature to which humid air must be cooled at constant pressure to become saturated.
3—The actual amount of the water vapor in a mixture of air and water.
4—The ratio of the water vapor actually present in the atmosphere to the amount that would be present if the air were saturated at the prevailing temperature and pressure.

5537. If the operating pressure is 2,000 PSI in a 1/2–inch line feeding an actuating cylinder which has a piston area of 10 square inches, the maximum weight the piston can lift will be

1—40,000 lb.
2—20,000 lb.
3—10,000 lb.
4—2,000 lb.

5538. Which of the following is the temperature to which humid air must be cooled at constant pressure to become saturated?

1—Dew point.
2—Vapor pressure.
3—Absolute humidity.
4—Relative humidity.

5539. If both the volume and the absolute temperature of a confined gas are doubled, the pressure will

1—not change.
2—be doubled.
3—be halved.
4—become four times as great.

5540. FAA Airworthiness Directives are issued to

1—provide temporary maintenance procedure.
2—prescribe airman privileges and limitations.
3—present suggested maintenance procedures.
4—correct an unsafe condition.

5541. (1) A Supplemental Type Certificate may be issued to more than one applicant for the same design change, providing each applicant shows compliance with the applicable airworthiness requirement.

(2) An installation of an item manufactured in accordance with the Technical Standard Order system requires no further approval for installation in a particular aircraft.

Regarding the above statements, which of the following is true?

1—Only No. 2 is true.
2—Both No. 1 and No. 2 are true.
3—Neither No. 1 nor No. 2 is true.
4—Only No. 1 is true.

5542. For all aircraft type certificated prior to January 1, 1958, the CAA issued documents known as Aircraft Specifications which contained technical information about the aircraft type. If an aircraft was type certificated May 7, 1959, the information that would formerly have been contained in the Aircraft Specifications will be in the appropriate

1—Aircraft Operation Information Letters.
2—Certificated Aircraft Bulletins.
3—Type Certificate Data Sheets.
4—Aviation Airworthiness Alerts.

5543. Which of the following issues Airworthiness Directives?

1—National Transportation Safety Board.
2—Air Transport Association.
3—Manufacturers.
4—Federal Aviation Administration.

5544. Which of the following is contained in a Type Certificate Data Sheet?

1—Maximum fuel grade to be used.
2—All pertinent minimum weights.
3—Control surface adjustment points.
4—Location of the datum.

5545. Suitability for use of a specific propeller with a particular engine–airplane combination can be determined by reference to what informational source?

1—Propeller Listing.
2—Propeller Specifications.
3—Aircraft Specifications or Type Certificate Data Sheet.
4—Alphabetical Index of Current Propeller Type Certificate Data Sheets, Specifications, and Listings.

5546. If an airworthy aircraft is sold, what is done with the Airworthiness Certificate?

1—It must be endorsed by a certificated mechanic to indicate that the aircraft is still airworthy.
2—It becomes invalid until the aircraft is reinspected and returned to service.
3—It is declared void and a new certificate is issued upon application by the new owner.
4—It is transferred with the aircraft.

5547. FAA Airworthiness Directives

1—are mandatory.
2—present design changes.
3—provide temporary maintenance procedures.
4—provide suggested maintenance procedures.

5548. Which of the following governs the issuance of an Airworthiness Certificate?

1—FAR Part 23.
2—FAR Part 21.
3—FAR Part 25.
4—FAR Part 39.

5549. Specifications pertaining to an aircraft, of which a limited number were manufactured under a type certificate and for which there is no current Aircraft Specification, can be found in the

1—Alphabetical Index of Antique Aircraft.
2—Aircraft Listing.
3—FAA Statistical Handbook of Civil Airplane Specifications.
4—Annual Summary of Deleted and Discontinued Aircraft Specifications.

5550. Where are technical descriptions of certificated propellers found?

1—Applicable Airworthiness Directives.
2—Aircraft Listings.
3—Aircraft Specifications.
4—Propeller type certificates.

5551. Which of the following is generally contained in Aircraft Specifications or Type Certificate Data Sheets?

1—Empty weight of the aircraft.
2—Useful load of aircraft.
3—Payload of aircraft.
4—Control surface movements.

5552. Placards required on an aircraft are specified in

1—Advisory Circular 43.13–1A.
2—the aircraft logbook.
3—Federal Aviation Regulations under which the aircraft was type certificated.
4—Aircraft Specifications or Type Certificate Data Sheets.

5553. Technical information about older aircraft models, of which no more than 50 remain in service, can be found in the

1—Aircraft Listing.
2—Annual Summary of Deleted and Discontinued Aircraft Specifications.
3—Alphabetical Index of Antique Aircraft.
4—FAA Statistical Handbook of Civil Airplane Specifications.

5554. (1) The Federal Aviation Regulations require approval after compliance with the data of a Supplemental Type Certificate.

(2) An installation of an item manufactured in accordance with the Technical Standard Order system requires no further approval for installation in a particular aircraft.

Regarding the above statements, which of the following is true?

1—Only No. 2 is true.
2—Both No. 1 and No. 2 are true.
3—Neither No. 1 nor No. 2 is true.
4—Only No. 1 is true.

5555. Which of the following would provide information regarding instrument range markings for an airplane certificated in the normal category?

1—FAR Part 21.
2—FAR Part 25.
3—FAR Part 91.
4—FAR Part 23.

5556. (1) Propellers are not included in the Airworthiness Directive system.

(2) A certificated powerplant mechanic may make a minor repair on a propeller and approve for return to service.

Regarding the above statements, which of the following is true?

1—Only No. 1 is true.
2—Only No. 2 is true.
3—Both No. 1 and No. 2 are true.
4—Neither No. 1 nor No. 2 is true.

5557. What publication would a mechanic use to determine the fuel quantity and octane rating for a specific aircraft?

1—Type Certificate Data Sheet.
2—Manufacturers' Parts Manual.
3—Summary of Supplemental Type Certificates.
4—Federal Aviation Regulations.

5558. An aircraft mechanic is privileged to perform major alterations on United States certificated aircraft; however, the work must be done in accordance with technical data approved by the Administrator before the aircraft can be approved for return to service. Which of the following is not approved data?

1—FAA Airworthiness Directives.
2—Advisory Circular 43.13–2A.
3—Type Certificate Data Sheets.
4—Manufacturers' manuals when FAA approved.

5559. What is the maintenance recording responsibility of the person who complies with an Airworthiness Directive?

1—Advise the aircraft owner/operator of the work performed.
2—Make an entry in the maintenance record of that equipment.
3—Advise the FAA district office of the work performed, by submitting an FAA Form 337.
4—Record of entry is not required of the person performing the work. The aircraft owner/operator is responsible for recording maintenance.

5560. (1) Manufacturer's data and FAA publications such as Airworthiness Directives, Type Certificate Data Sheets, and Advisory Circulars are all approved data.

(2) FAA publications such as Technical Standard Orders, Airworthiness Directives, Type Certificate Data Sheets, and Aircraft Specifications and Supplemental Type Certificates are all approved data.

In regard to the above statements, which is correct?

1—Both No. 1 and No. 2 are true.
2—Only No. 1 is true.
3—Neither No. 1 nor No. 2 is true.
4—Only No. 2 is true.

5561. The following is the compliance portion of an Airworthiness Directive. "Compliance required as indicated, unless already accomplished:

I. Aircraft with less than 500–hours' total time in service: Inspect in accordance with instructions below at 500–hours' total time, or within the next 50–hours' time in service after the effective date of this AD, and repeat after each subsequent 200 hours in service.

II. Aircraft with 500–hours' through 1,000–hours' total time in service: Inspect in accordance with instructions below within the next 50–hours' time in service after the effective date of this AD, and repeat after each subsequent 200 hours in service.

III. Aircraft with more than 1,000–hours' time in service: Inspect in accordance with instructions below within the next 25–hours' time in service after the effective date of this AD, and repeat after each subsequent 200 hours in service."

An aircraft has a total time in service of 468 hours. The Airworthiness Directive given was initially complied with at 454 hours in service. How many additional hours in service may be accumulated before the AD must again be complied with?

1—32.
2—46.
3—200.
4—186.

5562. Using Figure 51, which type of lubricant is used for the nose door link?

1—Oil, general purpose, low temperature lubricating.
2—Grease, general purpose.
3—Oil, lubricating, aircraft engine.
4—Grease, low temperature aircraft lubricating.

5563. Determine the frequency of lubrication of the pivot bushing. (See Figure 51.)

1—100 hours.
2—50 hours.
3—500 hours.
4—1,000 hours.

5564. Using Figure 51, what method is used to lubricate the nose door links?

1—Filler can.
2—Grease gun.
3—Hand.
4—Squirt can.

FREQUENCY	METHOD	TYPE OF LUBRICANT
⭘ 50 HOUR	🖐 HAND	GGP .... Grease, General Purpose
		GLT .... Grease, Low Temperature Aircraft Lubricating (Low Volatility Type)
⬜ 100 HOUR		OGP .... Oil, General Purpose, Low Temperature Lubricating
		OHA .... Hydraulic Fluid, Petroleum Base
		FG .... Graphite, Lubricating
◇ 500 HOUR	🛢 SQUIRT CAN	OAI .... Oil, Lubricating Aircraft Instrument (Low Volatility)
		P .... Petrolatum or Terminal Grease
⬡ 1000 HOUR	🛢 FILLER CAN	OEA .... Oil, Lubricating, Aircraft Engine. Aviation Grade Straight Mineral Oil SAE 50 above 40°F SAE 30 Below 40°F Grade 1100 above 40°F Grade 1065 below 40°F If detergent oil is used after the first 25 hours of operation, it must conform to Continental Motors Specification MHS-24A.
⬓ ON ASSEMBLY	🔫 GREASE GUN	Where OEA is shown with squirt can, use SAE 20.

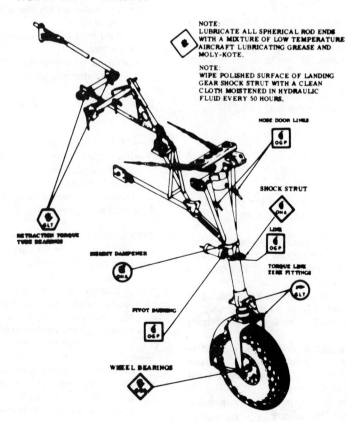

NOSE GEAR LINKAGE

NOTE:
LUBRICATE ALL SPHERICAL ROD ENDS WITH A MIXTURE OF LOW TEMPERATURE AIRCRAFT LUBRICATING GREASE AND MOLY-KOTE.

NOTE:
WIPE POLISHED SURFACE OF LANDING GEAR SHOCK STRUT WITH A CLEAN CLOTH MOISTENED IN HYDRAULIC FLUID EVERY 50 HOURS.

NOSE DOOR LINKS

SHOCK STRUT

RETRACTION TORQUE TUBE BEARINGS

LINK

SHIMMY DAMPENER

TORQUE LINK ZERK FITTINGS

PIVOT BUSHING

WHEEL BEARINGS

Figure 51

5565. The following is a table of airspeed limits as given in an FAA–issued aircraft specification:

Normal operating speed......................................260 knots
Never–exceed speed...293 knots
Maximum landing gear
   operation speed..174 knots
Maximum flap extended speed...........................139 knots

The high end of the white arc on the airspeed instrument would be at

1—260 knots.
2—293 knots.
3—174 knots.
4—139 knots.

5566. A complete detailed inspection and adjustment of the valve mechanism will be made at the first 25 hours after the engine has been placed in service. Subsequent inspections of the valve mechanism will be made each second 50–hour period.

From the above statement, at what intervals will valve mechanism inspections be performed?

1—100 hours.
2—25 hours.
3—50 hours.
4—75 hours.

5567. Check thrust bearing nuts for tightness on new or newly overhauled engines at the first 50–hour inspection following installation. Subsequent inspections on thrust bearing nuts will be made at each third 50–hour inspection.

From the above statement, at what intervals should you check the thrust bearing nut for tightness?

1—150 hours.
2—50 hours.
3—100 hours.
4—200 hours.

5568. Which statement is true regarding the privileges of a certificated mechanic with a powerplant rating?

1—They may perform the 100–hour inspection required by the Federal Aviation Regulations on a powerplant or propeller or any component thereof, but may not release the same to service.
2—They may perform the annual inspection required by the Federal Aviation Regulations on a powerplant or propeller or any component thereof, and may release the same to service.
3—They may perform the annual inspection required by the Federal Aviation Regulations on airframe, powerplant, or propeller or any component thereof, and may release the same to service.
4—They may perform the 100–hour inspection required by the Federal Aviation Regulations on a powerplant or propeller or any component thereof, and may release the same to service.

5569. Who has the authority to approve an aircraft for return to service after a 100–hour inspection?

1—A mechanic of any certificated repair station.
2—A certificated mechanic with an airframe rating.
3—A mechanic holding a repairman certificate.
4—A certificated mechanic with an airframe and powerplant rating.

5570. A repair, as performed on an airframe, shall mean

1—the upkeep and preservation of the airframe including the component parts thereof.
2—the restoration of the airframe to a condition for safe operation after damage or deterioration.
3—an appreciable change in the weight, balance, structural strength, performance, flight characteristics, or other qualities affecting the airworthiness of the airframe.
4—simple or minor preservation operations and the replacement of small standard parts not involving complex assembly operations.

5571. The replacement of fabric on fabric–covered parts such as wings, fuselages, stabilizers, or control surfaces is considered to be a

1—minor repair unless the new cover is different in any way from the original cover.
2—minor repair unless the underlying structure is altered or repaired.
3—major repair even though no other alteration or repair is performed.
4—minor repair unless the method of covering is changed or the underlying structure is altered or repaired.

5572. Which of the following is classified as a major repair?

1—Removal, installation, and repair of landing gear tires.
2—The repair of portions of skin sheets by making additional seams.
3—Troubleshooting and repairing broken circuits in landing light wiring circuits.
4—Replacing safety belts.

5573. The 100–hour inspection required by Federal Aviation Regulations for certain aircraft being operated for hire may be performed by

1—a person working under the supervision of an appropriately rated mechanic, but the aircraft must be approved by the mechanic for return to service.
2—an appropriately rated mechanic only if he/she has an inspection authorization.
3—an appropriately rated mechanic and approved by him/her for return to service.
4—an appropriately rated mechanic, but the aircraft must be approved for return to service by a mechanic with an inspection authorization.

5574. A person working under the supervision of a certificated mechanic with an airframe and powerplant rating is not authorized to perform which of the following?

1—Repair fabric covering involving an area greater than that required to repair two adjacent wing ribs.
2—Repair a wing brace strut by welding.
3—A 100–hour inspection.
4—Repair an engine mount by riveting.

5575. A mechanic certificate is effective for what period of time?

1—As long as duties are performed under the certificate for at least 60 days within the preceding 6 months.
2—24 calendar months.
3—As long as duties are performed under the certificate for at least 90 days within the preceding 12 months.
4—Until surrendered, suspended, or revoked.

5576. A certificated mechanic is not privileged to perform any repair to, or alteration of

1—a propeller.
2—air conditioning and pressurization systems.
3—an electrically heated windshield system.
4—an instrument.

5577. An Airworthiness Directive requires that a propeller be altered. A certificated mechanic could

1—perform and approve the work for return to service if it is a minor alteration.
2—not approve the work for return to service because it is an alteration.
3—not perform the work because it is an alteration.
4—not perform the work because he/she is not allowed to perform and approve for return to service, repairs or alterations to propellers.

5578. Which of the following is a privilege of a certificated mechanic with a powerplant rating?

1—Minor repairs to, and minor alterations of, propellers.
2—Minor repairs to, and major alterations of, propellers.
3—Major repairs to, and minor alterations of, propellers.
4—Major repairs to, and major alterations of, propellers.

5579. FAA certificated mechanics may

1—approve for return to service a major repair for which they are rated.
2—supervise and approve a 100–hour inspection.
3—perform an annual inspection appropriate to the rating(s) they hold.
4—approve for return to service a minor alteration they have performed appropriate to the rating(s) they hold.

5580. A certificated mechanic with a powerplant rating may perform the

1—annual inspection required by the Federal Aviation Regulations on a powerplant or any component thereof and approve and return the same to service.
2—100–hour inspection required by the Federal Aviation Regulations on a powerplant or any component thereof and approve and return the same to service.
3—annual inspection required by the Federal Aviation Regulations on an airframe, powerplant, or any component thereof and return the same to service.
4—100–hour inspection required by the Federal Aviation Regulation on an airframe, powerplant, or any other component thereof and approve and return the same to service.

5581. A certificated mechanic shall not exercise the privileges of the certificate and rating unless, within the preceding 24 months, the Administrator has found that the certificate holder is able to do the work or the certificate holder has

1—served as a mechanic under the certificate and rating for at least 18 months.
2—served as a mechanic under the certificate and rating for at least 12 months.
3—technically supervised other mechanics for at least 3 months.
4—served as a mechanic under the certificate and rating for at least 6 months.

5582. (1) A certificated mechanic with an airframe rating may perform a minor repair to an airspeed indicator providing he/she has the necessary equipment available.

(2) A certificated mechanic with a powerplant rating may perform a major repair to a propeller providing he/she has the necessary equipment available.

Regarding the above statements, which of the following is true?

1—Only No. 1 is true.
2—Neither No. 1 nor No. 2 is true.
3—Only No. 2 is true.
4—Both No. 1 and No. 2 are true.

5583. A certificated mechanic must notify the Administrator of the Federal Aviation Administration of any changes in permanent mailing address in writing within

1—48 hours after such changes.
2—30 days after such changes.
3—60 days after such changes.
4—90 days after such changes.

5584. Who is responsible for determining that materials used in aircraft maintenance and repair are of the proper type and conform to the appropriate standards?

1—The installing person or agency.
2—The owner of the aircraft.
3—The supplier of the material.
4—The manufacturer of the aircraft.

5585. The replacement of a damaged engine mount with a new identical engine mount purchased from the aircraft manufacturer is considered a

1—major alteration.
2—minor alteration.
3—major repair.
4—minor repair.

5586. Who has the authority to approve for return to service a powerplant or propeller or any part thereof after a 100–hour inspection?

1—A mechanic with a powerplant rating.
2—A mechanic with an airframe rating.
3—Any certificated repairman.
4—Personnel of any certificated repair station.

5587. Instrument repairs may be performed

1—by the instrument manufacturer only.
2—by an FAA–approved instrument repair station.
3—on airframe instruments by mechanics with an airframe rating.
4—on powerplant instruments by mechanics with a powerplant rating.

GENERAL TEST BOOK
ANSWERS, EXPLANATIONS, AND REFERENCES

5001. ANSWER #2
The working voltage of the capacitor must be at least 50% greater than the applied voltage. If the voltage applied across the plates is too great, the dielectric will break down and arcing will occur between the plates. The capacitor is then short-circuited, and the possible flow of direct current through it can cause damage to other parts of the equipment.
REFERENCE: EA-ITP-GB, Page 159; EA-AC65-9A, Page 351

5002. ANSWER #4
When the capacitive reactance in a circuit is the greatest, the circuit is said to be capacitive.
REFERENCE: EA-ITP-GB, Page 170; EA-AC65-9A, Pages 354-355

5003. ANSWER #4
The opposition to the flow of alternating current caused by the generation of a back voltage as the magnetic flux cuts accross the conductor is called inductive reactance. Its symbol is X_L, and it is measured in ohms.
REFERENCE: EA-ITP-GB, Page 153; EA-AC65-9A, Page 345

5004. ANSWER #1
Inductive reactance is a function of the amount of inductance and the frequency of the circuit. It increases with an increase in both frequency and inductance.
REFERENCE: EA-ITP-GB, Page 153; EA-AC65-9A, Page 345

5005. ANSWER #1
The Ohm is the standard unit of resistance, or opposition to current flow. More practically, it is the resistance through which a force of one volt can force a flow of one ampere.
REFERENCE: EA-ITP-GB, Page 93; EA-AC65-9A, Page 273

5006. ANSWER #4
At some particular frequency, known as the resonant frequency, the reactive effects of a capacitor and an inductor will be equal. These effects are opposite of one another and will cancel, leaving only the Ohmic value of resistance to oppose current flow.
REFERENCE: EA-ITP-GB, Page 175; EA-AC65-9A, Page 357

5007. ANSWER #4
The effective value (voltage) in an AC circuit is .707 times the RMS (root-mean-square) of the instantaneous voltage.
REFERENCE: EA-ITP-GB, Page 146; EA-AC65-9A, Page 343

5008. ANSWER #3
The capacity of a capacitor is affected by three variables: the area of the plates, the separation between the plates, and the dielectric constant of the material between the plates. It is directly proportional to the plate area, and inversely proportional to the distance between the plates.
REFERENCE: EA-ITP-GB, Page 158; EA-AC65-9A, Page 346

5009. ANSWER #1
A transformer does not generate any power. Realistically, due to heat, it loses power. Because this question says that there are no losses, the voltage in the secondary being stepped up five times would mean that the amperage in the secondary would only be one-fifth; 0.02 amps in the secondary would mean that there must be one amp in the primary.
REFERENCE: EA-ITP-GB, Page 155; EA-AC65-9A, Page 359

5010. ANSWER #2
In a purely inductive circuit, the current will not begin to flow until the voltage has risen to its peak value. As the voltage begins to drop off, the current rises until the voltage reaches zero. The current in a circuit like this is lagging the voltage by 90°.
REFERENCE: EA-ITP-GB, Page 139; EA-AC65-9A, Page 352

5011. ANSWER #1
When one coulomb flows past a point in one second, there is a flow of one ampere, or one amp.
REFERENCE: EA-ITP-GB, Page 93; EA-AC65-9A, Page 272

5012. ANSWER: #3
There are three values of alternating current. They are instantaneous, maximum, and effective. Unless specified otherwise, all current and voltage references are considered to be effective values.
REFERENCE: EA-AC65-9A, Page 343

5013. ANSWER #3
Each lamp requires 72 Watts of power (3 amps × 24 volts = 72 Watts), or a total of 144 Watts. The other choices would require 96 Watts, 120 Watts, or 99.46 Watts.
REFERENCE: EA-ITP-GB, Pages 107-108; EA-AC65-9A, Page 284

5014. ANSWER #4
Power is equal to system voltage multiplied by each item's current flow (P = IE). The motor requires 198.9 Watts, the position lights 60 Watts, the heating element 120 Watts, and the anticollision light 72 Watts.
REFERENCE: EA-ITP-GB, Pages 107-108; EA-AC65-9A, Page 284

5015. ANSWER #1
To find the amps, divide Watts (746 Watts = 1 HP) by three. This will give you the Watts in 1/3 HP. The result, 248.66 Watts, divided by the voltage (24v) will give you the amps, 10.366. Rounded to 10.4, the choice is #1.
REFERENCE: EA-ITP-GB, PAGE 109, EA-AC65-9A, Page 286

5016. ANSWER #3
To answer this question you must first find out what the current flow is for the unit. Current equals system voltage divided by resistance. For this unit, current (I) = 28 ÷ by 10, or 2.8 amps. Power = volts × amps (P = IE), or 28 × 2.8.
REFERENCE: EA-ITP-GB, Pages 107-108; EA-AC65-9A, Pages 282-284

5017. ANSWER #1
If the efficiency of a motor remains the same, a one horsepower motor operating under any system voltage (12, 24, etc.) will require the same number of Watts of power. The advantage of the higher voltage motor is that its current draw is less, and could operate with a smaller size wire (feed line).
REFERENCE: EA-ITP-GB, Pages 107-108; EA-AC65-9A, Page 284

5018. ANSWER #4
To solve this problem, it is first necessary to find the total resistance of the circuit. In a parallel circuit, the total resistance is equal to:

$$\text{Resistance Total} = \frac{1}{\dfrac{1}{R_1} + \dfrac{1}{R_2} + \dfrac{1}{R_3}} \quad \text{(FOR THE NUMBER OF RESISTORS)}$$

The total resistance for this circuit is 1.11 Ohms. This resistance divided into the system voltage gives a current flow of 25.23 amps.
REFERENCE: EA-ITP-GB, Pages 107-108; EA-AC65-9A, Pages 282-284

5019. ANSWER #1
Horsepower is doing 33,000 foot-pounds of work in one minute. This is a *rate of work* reference and is the standard reference.
REFERENCE: EA-ITP-GB, Page108; EA-AC65-9A, Page 285

5020. ANSWER #3
With carbon resistors, the larger the physical size, the more power they are able to dissipate.
REFERENCE: EA-ITP-GB, Page 113

5021. ANSWER #2
Potential is just another way of talking about the unit of electrical pressure known as voltage.
REFERENCE: EA-ITP-GB, Page 93; EA-AC65-9A, Page 271

5022. ANSWER #3
The apparent power is equal to the true power divided by the power factor. Because the power factor is given as a percent, for example 50%, it is necessary to multiply the true power by 100.
REFERENCE: EA-ITP-GB, Pages 142-143; EA-AC65-9A, Pages 357-358

5023. ANSWER #4
In a parallel circuit, the voltage supplied to each resistor is equal to system voltage. Across each of the resistors, all of the voltage is dropped (in this case, a 24 volt drop across each resistor).
REFERENCE: EA-ITP-GB, Page 119; EA-AC65-9A, Pages 357-358

5024. ANSWER #4
In an alternating current circuit, the current is not always in phase with the voltage. For this reason, the apparent power, or what would appear to be present, is more than the true power available to the circuit.
REFERENCE: EA-ITP-GB, Pages 142-143; EA-AC65-9A, Pages 357-358

5025. ANSWER #3
Power is equal to volts multiplied by amps. In the circuit, the current flow would be 2 amps (volts divided by resistance) multiplied by the system voltage of 24.
REFERENCE: EA-ITP-GB, Pages 107-108; EA-AC65-9A, Pages 282-284

5026. ANSWER #3
With resistor R_5 disconnected, the Ohmmeter will read the resistance of the other four resistors. Resistors R_3 and R_4 are in series with each other, for a total resistance of 12 Ohms. This 12 Ohm total is in parallel with the other 2 resistors. By using the formula given in the explanation for question #5018, the total resistance is found to be 3 Ohms.
REFERENCE: EA-ITP-GB, Pages 107-108; EA-AC65-9A, Pages 282-284

5027. ANSWER #4
An Ohmmeter measures resistance, therefore it must have power in order to make the measurement. A small flashlight or penlight battery is commonly mounted in the meter case.
REFERENCE: EA-ITP-GB, Page 208; EA-AC65-9A Page 330

5028. ANSWER #1
A D'Arsonval meter is a current (amperage) measuring instrument. With the proper sensitivity, the meter may be used in a circuit without any additional components. If the range of current to be measured is greater than the meter scale, a shunt must be installed in parallel with the meter.
REFERENCE: EA-ITP-GB Page 206, EA-AC65-9A Page 322

5029. ANSWER #1
By disconnecting resistor R_3 at terminal D, the path of flow is broken and the resistance read by the ohmmeter would be infinite.
REFERENCE: EA-ITP-GB, Page 116; EA-AC65-9A, Page 276

5030. ANSWER #4
Because of the break in the circuit, resistors R_1 and R_2 are the only ones that will be read by the Ohmmeter. These two resistors are of equal value, and they are in parallel. When parallel resistors are of equal value the total resistance is equal to the value of any one of the resistors divided by the number of resistors. In this case, it would be 20 divided by 2, or 10 Ohms.
REFERENCE: EA-ITP-GB, Page 119; EA-AC65-9A, Page 292

5031. ANSWER #3
The voltmeter in parallel with the light, the ammeter in series with the light and the battery, are connected properly. The polarity on the second voltmeter is incorrect and the second ammeter is in parallel with the battery, which would cause it to burn up.
REFERENCE: EA-AC65-9A, Pages 326-328

5032. ANSWER #1
The voltage available from the secondary of a transformer is dependent upon the number of turns in the secondary winding. The transformer in the engine's magneto uses many turns in the secondary winding to produce a high output voltage.
REFERENCE: EA-ITP-GB, Page 155; EA-AC65-9A, Page 359

5033. ANSWER #3
By connecting the voltmeter in parallel with the unit, you are able to read the voltage drop. This will tell you the voltage being supplied to the item.
REFERENCE: EA-AC65-9A, Pages 326-328

5034. ANSWER #3
In a parallel circuit, the voltage supplied to each item in the circuit is the same.
REFERENCE: EA-ITP-GB, Page 167; EA-AC65-9A, Pages 290-291

5035. ANSWER #3
The megger is often used for measuring insulation resistance in ignition systems and other high-voltage circuits.
REFERENCE: EA-ITP-GB, Pages 210-211; EA-AC65-9A, Page 331

5036. ANSWER #3
Continuity lights have their own source of power, so the circuit being checked should not be connected to any other source of power.
REFERENCE: EA-AC65-15A, Page 465

5037. ANSWER #2
Power is equal to the current squared (I^2) multiplied by the resistance. For this problem, the power would be 0.035 Watts, or 35 milliWatts.
REFERENCE: EA-ITP-GB, Page 108; EA-AC65-9A, Page 285

5038. ANSWER #1
Ideally the number of terminals on one stud should be limited to two. The maximum allowed is four.
REFERENCE: EA-ITP-AB, Page 340; EA-AC43.13-1A, Page 189

5039. ANSWER #4
There are four 1.5 volt batteries connected in series. When batteries are connected in series, their voltages are added.
REFERENCE: EA-AC65-9A, Pages 275, 310

5040. ANSWER #3
Power factor is a term used to indicate the amount of current that is in phase with the voltage, and it may be found as the ratio between the true power and the apparent power.
REFERENCE: EA-ITP-GB, Page 143; EA-AC65-9A, Pages 357-358

5041. ANSWER #4
Power, or Watts, is equal to the volts multiplied by the current. If there are 120 volts and 60 Watts of power, there must be ½ amp of current flow.
REFERENCE: EA-ITP-GB, Pages 107-108; EA-AC65-9A, Pages 283-284

5042. ANSWER #4
Power being equal to the voltage multiplied by the current, the anticollision light would require 144 Watts of power. The four lamps require only 120 Watts, the landing gear motor only 96 Watts, and the ¹/10 horsepower motor only 99.4 Watts.
REFERENCE: EA-ITP-GB, Pages 107-108; EA-AC65-9A, Pages 283-284

5043. ANSWER #3
Power in electrical systems is measured in Watts.
REFERENCE: EA-ITP-GB, Page 93; EA-AC65-9A, Page 284

5044. ANSWER #3
Answering this question involves the formulas for Ohm's law and power. Current flow equals power divided by volts, so the current in this example is 1.07 amps. Resistance, according to Ohm's law, is equal to volts divided by amps. Twenty-eight volts divided by 1.07 amps equals 26 Ohms.
REFERENCE: EA-ITP-GB, Pages 107-108; EA-AC65-9A, Pages 283-284

5045. ANSWER #3
In a parallel circuit, each of the units in the circuit has system voltage available to it. The current flow through each unit, however, is dependent on the unit's resistance. By totalling up the individual current flows in the circuit, you can get the total current flow.
REFERENCE: EA-ITP-GB, Page 119; EA-AC65-9A, Pages 290-291

5046. ANSWER #3
Diodes are current blocking devices. By installing one in an AC electrical circuit, one direction of current flow is blocked. This changes the alternating current into direct current, or rectifies it.
REFERENCE: EA-ITP-GB, Page 181; EA-AC65-9A, Pages 373-374

5047. ANSWER #4
In a series circuit, current flow is constant. Current is equal to the circuit voltage divided by the total resistance. Because resistance values are added in a series circuit, the total resistance here is 30 Ohms. Circuit voltage of 28 divided by total resistance of 30 gives a current flow of 0.93 amps.
REFERENCE: EA-ITP-GB, Pages 116-117; EA-AC65-9A, Pages 286-287

5048. ANSWER #1
A good conductor is one which offers the least resistance to current flow.
REFERENCE: EA-ITP-GB, Page 109; EA-AC65-9A, Page 273

5049. ANSWER #2
In a series circuit, current is constant and voltage varies (decreases) as it pushes the current through each of the resisitors. The voltage drop in each resistor is equal to the current multiplied by the resistance. In this case it would be 1 amp multiplied by 10 Ohms, or a voltage drop of 10.
REFERENCE: EA-ITP-GB, Page 117; EA-AC65-9A, Page 288

5050. ANSWER #3
Current flow in a parallel circuit varies, according to the resistors in the circuit. Current flowing through C-D in the circuit is on its way to resistors R_2 and R_3, which are in parallel. Their combined resistance is 8 Ohms, so 24 volts divided by 8 Ohms gives a current flow of 3 amps.
REFERENCE: EA-ITP-GB, Pages 118-119; EA-AC65-9A, Pages 290-293

5051. ANSWER #4
Voltage in a parallel circuit is constant to each of the resistors. Voltage will drop completely across each of the resistors, so the voltage drop across any one of the resistors in this circuit would be 24 volts.
REFERENCE: EA-ITP-GB, Pages 118-119; EA-AC65-9A, Pages 290-293

5052. ANSWER #4
Resistors 4 and 5 in this circuit are parallel to each other, giving a total resistance of 4 Ohms. This 4 Ohms is in series with resistor 2, giving a total resistance of 16 Ohms. This 16 Ohms is in parallel with resistor 3, giving a total of 3.2 Ohms. This 3.2 Ohms is in series with resistor 1, giving a circuit total of 21.2 Ohms.
REFERENCE: EA-ITP-GB, Pages 120-122; EA-AC65-9A, Pages 292-294

5053. ANSWER #2
In a parallel circuit, the greater the number of resistors, the less the total resistance. If you remove a resistor from the circuit, the total resistance in the circuit goes up. It is for this reason that plugging too many appliances into a single outlet in your house will cause the circuit breaker to pop. The total resistance in the circuit goes down, the current flow goes up, and the breaker is overloaded.
REFERENCE: EA-ITP-GB, Chapter 2, Pages 30-31; EA-AC65-9A, Pages 290-291

5054. ANSWER #1

Current flow is equal to the voltage divided by the resistance. If 25 amps are flowing at 110 volts, the resistance must be 4.4 Ohms. If the voltage is decreased to 85 volts, the 85 divided by 4.4 would give a current flow of 19.3 amps.
REFERENCE: EA-ITP-GB, Page 107; EA-AC65-9A, Pages 282-283

5055. ANSWER #2

The total current flow in the circuit is equal to the sum of the currents through each of the resistors. Resistor 1 has .4 amps of current (12 divided by 30), resistor 2 has .2 amps of current (12 divided by 60), and resistor 3 has .8 amps of current (12 divided by 15). This gives a total current flow of 1.4 amps.
REFERENCE: EA-ITP-GB, Pages 118-119; EA-AC65-9A, Pages290-291

5056. ANSWER #4

Resistors 2, 3, and 4 are in parallel with each other. Their combined resistance is 2 Ohms. This combined resistance is in series with resistors 1 and 5, giving a total resistance of 17 Ohms.
REFERENCE: EA-ITP-GB, Chapter 2, Pages 30-31; EA-AC65-9A, Pages 290-291

5057. ANSWER #2

Magnetic lines of flux pass most easily through ferrous materials, which include iron and iron alloys. The other selections for this question are non-ferrous metals.
REFERENCE: EA-ITP-GB Pages 98-99; EA-AC65-9A, Page 298

5058. ANSWER #1

The power in a circuit, or Watts, is equal to the circuit voltage multiplied by the current flow. In this circuit, there would need to be 4 amps of current flow to produce 192 Watts with a voltage of 48. Using Ohm's law, 48 volts and 4 amps would mean there must be a total circuit resistance of 12 Ohms. Three resistors of 36 Ohms each, in parallel, would give a total resistance of 12 Ohms.
REFERENCE: EA-ITP-GB, Pages 107-108 and 118-119; EA-AC65-9A, Pages 284-285, 290-291

5059. ANSWER #1

In a parallel circuit, as resistors are added, the total resistance of the circuit decreases. The value of any one resistor will always be greater than the total resistance of the circuit.
REFERENCE: EA-ITP-GB, Pages 118-119; EA-AC65-9A, Pages 290-291

5060. ANSWER #4

In a series circuit, the current flow multiplied by the resistance gives the voltage drop across that resistor. In a parallel circuit, the current flow multiplied by the resistance gives not only the voltage drop across the resistor, but also the voltage in the circuit (since voltage to each resistor is the same).
REFERENCE: EA-ITP-GB, Pages 107-108; EA-AC65-9A, Pages 284-285

5061. ANSWER #2

Thermal switches are designed to open the circuit automatically whenever the temperature of a motor becomes excessively high.
REFERENCE: EA-ITP-GB, Page 113; EA-AC65-9A, Page 317

5062. ANSWER #1

Through the up limit switch, wire #19 supplies power to the red light. With an open in this wire, the only way to get power to the light is with the push-to-test circuit.
REFERENCE: EA-ITP-GB, Pages 215-219; EA-AC65-9A, Pages 332-338

5063. ANSWER #2

The number 7 wire is the power source from the 5 amp circuit breaker which feeds power directly to the red and green lights. The circuit is only completed, however, when a path to the ground is given by the push-to-test.
REFERENCE: EA-ITP-GB, Pages 215-219; EA-AC65-9A, Pages332-338

5064. ANSWER #4

If the PCO relay does not operate, there is no way for switch #13 to close. This switch is what allows the 24 VDC bus to power, through the 5 amp breaker, the fuel pressure cross feed valve open "light".
REFERENCE: EA-ITP-GB, Pages 215-219; EA-AC65-9A, Pages 332-338

5065. ANSWER #2

The fuel selector switch in the cross-feed position powers the FCF relay, which in turn powers switch number 17. Through this switch the cross-feed valve is energized, closing switch 19 and allowing relay TCO to be energized.
REFERENCE: EA-ITP-GB, Pages 215-219; EA-AC65-9A, Pages 332-338

5066. ANSWER #1

When the system has power to the bus, and the fuel selector is switched to the right-hand tank, power is fed to the TRS relay. This relay opens switch #7 and closes switch #8. Switch #7 opening causes the cross-feed valve switches 11 and 12 to change position. The change in position of switch #11 takes power away from relay PCC. The change in position of switch #12 feeds power to relay PCO. A total of 3 relays have been operated.
REFERENCE: EA-ITP-GB, Pages 215-219; EA-AC65-9A, Pages 332-338

5067. ANSWER #2
All relays are said to be spring loaded to the position shown. Power supplied to the bus would have a straight shot through switches 5, 7, 9, and 11 to the PCC relay. Power would also have a straight shot to the TCC relay through switches 18 and 20.
REFERENCE: EA-ITP-GB, Pages 215-219; EA-AC65-9A, Pages 332-338

5068. ANSWER #3
This question can be answered by elimination, as well as by careful examination of how the circuit is described. The first thing to realize is that switches 5/6, 7/8 and 9/10 must change positions together. Only answer #3 does not violate this fact. The left hand position energizes relay LTS, which changes the position of switches 5 and 6. Opening switch 5 takes power away from relay PCC allowing switch 15 to close (remember that switch 15 is shown closed right now because that is the spring loaded position). Taking power away from switch 11 and supplying power to switch 12 energizes the cross-feed valve and these switch positions change. The change in switch 12's position energizes relay PCO, which closes switch 13.
REFERENCE: EA-ITP-GB, Page 215-219; EA-AC65-9A, Pages 332-338

5069. ANSWER #3
This is the accepted symbol for a potentiometer.
REFERENCE: EA-AC65-9A, Page 52

5070. ANSWER #4
This is the accepted symbol for a variable capacitor.
REFERENCE: EA-ITP-AB, Page 334

5071. ANSWER #1
Wire number 4 is the one that provides a path to ground through the throttle switch and the left gear switch.
REFERENCE: EA-ITP-GB, Pages 215-219; EA-AC65-9A, Pages 332-338

5072. ANSWER #3
If the control valve switch were not in the neutral position, the warning horn would have a path to ground through wires 6, 5, 10, 11, 13, and 14. When the gear is down and the throttle is retarded, you don't want the horn to sound.
REFERENCE: EA-ITP-GB, Pages 215-219; EA-AC65-9A, Pages 332-338

5073. ANSWER #4
The only way for both gear switches to be in the path to ground is to have the left gear up and the right gear down.
REFERENCE: EA-ITP-GB, Pages 215-219; EA-AC65-9A, Pages 332-338

5074. ANSWER #1
Wire number 5 connects the left gear switch and the right gear switch. If this wire had an open, there would not be a path to ground.
REFERENCE: EA-ITP-GB, Pages 215-219; EA-AC65-9A, Pages 332-338

5075. ANSWER #4
With the landing gear up, wire number 6 is the one that provides a path to ground, through the right gear switch.
REFERENCE: EA-ITP-GB, Pages 215-219; EA-AC65-9A, Pages 332-338

5076. ANSWER #3
Schematic diagrams are good for troubleshooting because they show the flow within the system. They indicate the relationship of one component to the others in the system, but they do not show the component's location in the aircraft.
REFERENCE: EA-ITP-GB, Page 230; EA-AC65-9A, Pages 46-49

5077. ANSWER #1
The common reference point in a circuit is called the ground. This is the reference point from which most circuit voltages are measured. This point is normally considered to be at zero potential.
REFERENCE: EA-AC65-9A, Page 277

5078. ANSWER #4
When the voltmeter is connected across the open resistor, the voltmeter has closed the circuit by shunting (paralleling) the burned-out resistor, allowing current to flow. The resistance of the voltmeter, however, is so high that only a very small current flows in the circuit. The current is too small to light the lamp, but the voltmeter will read the battery voltage.
REFERENCE: EA-AC65-9A, Pages 332-333

5079. ANSWER #1
In the process of delivering 2 amps of current flow to a load, the battery is dropping from 12.6 volts to 10 volts. A decrease of 2.6 volts with a 2 amp current flow would be caused by a battery internal resistance of 1.3 ohms (2.6 volts divided by 2 amps equal 1.3 ohms).
REFERENCE: EA-ITP-GB, Pages 107-108; EA-AC65-9A, Pages 282-284

5080. ANSWER #2
The circuit which powers the green light (gear down and locked indicator light) consists of wires 6, 5, 4, and 3. This circuit ties together the nose gear, left gear, and right gear "down" switches. An open in either one would not allow the light to come on.
REFERENCE: EA-ITP-GB, Pages 215-219; EA-AC65-9A, Pages 332-338

5081. ANSWER #1
This is the accepted symbol for a variable resistor (also called a rheostat).
REFERENCE: EA-AC65-9A, Pages 52, 296

5082. ANSWER #1
Sodium bicarbonate should be applied, in solution form, to an area where lead-acid battery electrolyte has been spilled. The area must then be washed down with water.
REFERENCE: EA-ITP-GB, Page 132; EA-AC43.13-1A, Page 295

5083. ANSWER #2
When a lead-acid battery is fully charged, the electrolyte (sulfuric acid) is almost totally in solution. Because sulfuric acid has a very low freezing point, when it is mixed with water, the freezing point of the solution is very low. For example, the freezing point of the solution when the specific gravity is 1.300 is a minus 90°F.
REFERENCE: EA-ITP-GB, Page 131

5084. ANSWER #3
The low voltage of a discharged battery will allow a large amount of current to flow into it when the charge is first begun, but as the charge continues and the voltage rises, the current will decrease.
REFERENCE: EA-ITP-GB, Page 133; EA-AC65-9A, Page 313

5085. ANSWER #1
One of the nice features of a constant-current charger is that it can be used to charge batteries of different voltage at the same time. The batteries are connected in series so the current supply will be the same to each of them (current is constant in a series circuit).
REFERENCE: Aircraft Electricity and Electronics, Page 84

5086. ANSWER #1
The capacity of a battery is its ability to produce a given amount of current for a specified time, and is expressed in ampere-hours. One ampere-hour of capacity is the amount of electricity that is put into or taken from a battery when a current of one ampere flows for one hour. Seven amps flowing for five hours would equal thirty-five amp-hours of capacity.
REFERENCE: EA-ITP-GB, Page 131; EA-AC65-9A, Page 310

5087. ANSWER #3
When batteries are connected in parallel, their output voltage remains the same (assuming they are batteries of equal voltage). The amount of current flow, and therefore power, is increased in proportion to the number of batteries.
REFERENCE: EA-AC65-9A, Page 276

5088. ANSWER #4
If a nickel-cadmium battery has been properly serviced and is in good condition, each cell should have a voltage of between 1.55 and 1.80.
REFERENCE: EA-ITP-GB, Page 136

5089. ANSWER #1
A nickel-cadmium battery typically uses a 30%, by weight, solution of potassium hydroxide and water as the electrolyte.
REFERENCE: EA-ITP-GB, Page 135; EA-AC65-9A, Page 313

5090. ANSWER #2
Aircraft batteries are rated according to their voltage and ampere-hour capacity.
REFERENCE: EA-ITP-GB, Pages 130-131; EA-AC65-9A, Page 310

5091. ANSWER #1
The specific gravity of the electrolyte in a nickel-cadmium battery runs between 1.24 and 1.30 at 80° Fahrenheit. This specific gravity does not change as the battery is charged or discharged because the electrolyte acts only as a conductor.
REFERENCE: EA-ITP-GB, Page 135

5092. ANSWER #4
If the cell link connections are not tight enough, arcing and overheating can occur, which will leave blue marks on the hardware.
REFERENCE: EA-AC65-9A, Page 316

5093. ANSWER #1
If a nickel-cadmium battery is subjected to a high charge current, high temperature, or high liquid level, excessive spewing may result. Evidence of this is seen as white crystal deposits on the cells.
REFERENCE: EA-AC65-9A, Page 316

5094. ANSWER #4
The electrolyte level in a nickel-cadmium battery is at its highest immediately after the charging cycle has ended. If the level is not checked and adjusted at this time, it will drop within a few hours and be more difficult to properly service.
REFERENCE: EA-ITP-GB, Page 136; EA-AC65-9A, Page 314

5095. ANSWER #3
The electrolyte in a nickel-cadmium battery is at its lowest when it is discharged.
REFERENCE: EA-ITP-GB, Page 136; EA-AC65-9A, Page 314

5096. ANSWER #2
A cell in a fully charged nickel-cadmium battery is typically at 1.5 volts.
REFERENCE: EA-ITP-GB, Page 136

5097. ANSWER #4
When a nickel-cadmium battery is stored for a long period of time, it loses its charge. When this happens, the electrolyte is absorbed in the plates and the electrolyte level in the cell drops.
REFERENCE: EA-ITP-GB, Page 136; EA-AC65-9A, Page 314

5098. ANSWER #2
Because the voltage and the specific gravity of a nickel-cadmium battery remain relatively constant during the battery's various states of charge, the only way of actually determining the state-of-charge is to perform a discharge at a specified rate.
REFERENCE: EA-ITP-GB, Page 135

5099. ANSWER #4
Because the electrolyte level is lower when the battery is not fully charged, adding water at this time will cause the battery to be overserviced.
REFERENCE: EA-AC65-9A, Page 316

5100. ANSWER #3
The problem described in this question can be detected by the presence of white crystals on the cells.
REFERENCE: EA-AC65-9A, Page 316

5101. ANSWER #1
During charging, the negative plates lose oxygen and begin forming metallic cadmium. When all the oxygen is removed from the negative plates and only cadmium remains, the cells will start to gas.
REFERENCE: EA-ITP-GB, Page 135; EA-AC65-9A, Page 314

5102. ANSWER #1
Lines of this type are known as hidden lines.
REFERENCE: EA-ITP-GB, Page 233; EA-AC65-9A, Pages 44-45

5103. ANSWER #1
This line is made up of alternate long and short dashes.
REFERENCE: EA-ITP-GB, Page 233; EA-AC65-9A, Pages 43-44

5104. ANSWER #3
In the isometric view of the balance weight, the first thing to realize is that the back side of the weight is open to the internal cavity. This means that the center box representing that opening, when viewed from the back, will be made up of solid lines. The four openings coming in from the sides will not be visible when viewed from the back, so they will appear as dashes (hidden lines). The pointed end of the balance weight has two flats which will also show up as dashes (hidden lines).
REFERENCE: EA-ITP-GB, Pages 232-234; EA-AC65-9A, Pages 44-45

5105. ANSWER #2
Schematic diagrams are used for troubleshooting and for tracing the flow within a system. They do not identify locations within the aircraft.
REFERENCE: EA-ITP-GB, Page 230; EA-AC65-9A, Pages to 49

5106. ANSWER #4
Detail drawings describe a single part, and usually give information about its size, shape, material and method of manufacture. Assembly drawings describe an object made up of two or more parts.
REFERENCE: EA-ITP-GB, Pages 221-223; EA-AC65-9A, Page 35

5107. ANSWER #2
This is an example of an orthographic projection. The bottom view is obtained by rotating the object 90°, so the line that now forms the bottom of the object would form the top of the object. By doing this, the four sides would be seen (solid lines) and the setback in the object would not be seen (dashed line on the right side).
REFERENCE: EA-ITP-GB, Page 231; EA-AC65-9A, Pages 39-40

5108. ANSWER #1
A fuselage station number is used to identify the location of items in the aircraft. In weight and balance calculations, this location is called the "arm".
REFERENCE: EA-ITP-GB, Pages 230-239; EA-AC65-9A, Page 38

5109. ANSWER #1
These lines are generally drawn as thin, full lines, but they may vary with the kind of material being shown.
REFERENCE: EA-ITP-GB, Pages 233-234; EA-AC65-9A, Pages 44-45

5110. ANSWER #2
The six possible sides in an orthographic projection are the front, back, top, bottom, left side, and right side. The front is the principal view.
REFERENCE: EA-ITP-GB, Page 231; EA-AC65-9A, Pages 39-40

5111. ANSWER #4
As many as six views could be shown in an orthographic projection, but one, two or three is the most common.
REFERENCE: EA-ITP-GB, Page 231; EA-AC65-9A, Pages 39-40

5112. ANSWER #3
This is an example of an orthographic projection. The left side view is obtained by rotating the object 90°, so the line which now forms the left side of the object would be on the right of the side facing you. This would have you looking at four sides as visible (solid) lines, and a dashed line running horizontal where the setback is.
REFERENCE: EA-ITP-GB, Page 231; EA-AC65-9A, Pages 39-40

5113. ANSWER #1
The title block is usually in the lower right corner of a drawing, visible when the drawing is properly folded.
REFERENCE: EA-ITP-GB, Page 236; EA-AC65-9A, Page 36

5114. ANSWER #3
This line is shown as a series of short dashes.
REFERENCE: EA-ITP-GB, Page 233; EA-AC65-9A, Pages 44-45

5115. ANSWER #1
This is an example of an orthographic projection. The bottom view of the object is obtained by rotating the object 90°, so the line that now forms the bottom of the object would form the top of the object. By doing this, the four sides would be seen (solid lines) and the two setbacks would not be seen (vertical dashed lines).
REFERENCE: EA-ITP-GB, Page 231; EA-AC65-9A, Pages 39-40

5116. ANSWER #3
Schematic diagrams do not show the location of items in the aircraft, but they do show the relationship between components in the system.
REFERENCE: EA-ITP-GB, Page 230; EA-AC65-9A, Pages 46-49

5117. ANSWER #3
Phantom lines indicate the alternate position of parts of the object or the relative position of a missing part.
REFERENCE: EA-ITP-GB, Pages 233-234; EA-AC65-9A, Pages 44-45

5118. ANSWER #4
Aircraft maintenance technicians are always involved in removing and replacing parts. Installation diagrams are very helpful when the technician is performing this task.
REFERENCE: EA-ITP-GB, Page 224; EA-AC65-9A, Page 46

5119. ANSWER #1
The four steps in sketching are blocking in, adding detail, darkening the views, and then adding the dimensions.
REFERENCE: EA-AC65-9A, Page 49

5120. ANSWER #2
When making a sketch, the last step is to sketch the extension and dimension lines. This is the only step missing from the illustration.
REFERENCE: EA-AC65-9A, Page 49

5121. ANSWER #4
An orthographic projection contains as many as six views, but the use of three of the views (front, top, and right side) is the most common.
REFERENCE: EA-ITP-GB, Page 231; EA-AC65-9A, Pages 39-40

5122. ANSWER #2
If a sketch is going to be used to make a part, the sketch must contain all the information necessary for the manufacture. A sketch is not the best thing to use if a part is to be manufactured, but it is a possibility.
REFERENCE: EA-AC65-9A, Page 49

5123. ANSWER #3
The top, front, and right-side views are the ones normally portrayed in an orthographic projection.
REFERENCE: EA-ITP-GB, Page 231; EA-AC65-9A, Pages 39-40

5124. ANSWER #4
The first step in making a sketch is to block in the views. It is after the views are blocked in that the details are added.
REFERENCE: EA-AC65-9A, Page 49

5125. ANSWER #2
It is not necessary to include a detailed dimensional sketch of aircraft skin repairs in the permanent records. Sketches are more than likely going to made, but they are not included in the records.
REFERENCE: EA-AC65-9A, Page 49

5126. ANSWER #3
There are six accepted views in an orthographic projection.
REFERENCE: EA-ITP-GB, Page 231; EA-AC65-9A, Pages 39-40

5127. ANSWER #2
These three types of drawings are true working drawings. Detail drawings are for parts manufacturing, while the other two are self explanatory.
REFERENCE: EA-ITP-GB Pages 221-223; EA-AC65-9A Page 35

5128. ANSWER #1
Illustrating a repair or fabricating a replacement part are common uses of sketches, as they only need dimensions and need not be drawn to scale.
REFERENCE: EA-ITP-GB Page 244, EA-659A Pages 35-36

5129. ANSWER #1
A chamfer is a squared off corner of an object, or edge of an object, such as the round stock which this clevis pin is made of. In the answer to this question, the fraction 1/16 has been converted to its decimal equivalent 0.0625. The 1/16 is the length of the flat, and the 45° is the angle of the flat to the horizontal.
REFERENCE: EA-ITP-GB, Page 235 (implied, not specific); EA-AC65-9A, Page 43 (implied, not specific)

5130. ANSWER #3
This is an example of an orthographic projection. The bottom view of the object is obtained by rotating the object 90°, so the line that now forms the bottom of the object would form the top of the object. By doing this, the four sides would be seen (solid lines) and the two vertical pieces (made up of four lines) would not be seen (vertical dashed lines).
REFERENCE: EA-ITP-GB, Page 231; EA-AC65-9A, Pages 39-40

5131. ANSWER #1
When you put measurements on a drawing, you are said to be dimensioning the drawing.
REFERENCE: EA-ITP-GB, Page 233; EA-AC65-9A, Page 43

5132. ANSWER #1
Extension lines are the light lines that extend from the point at which a measurement is made.
REFERENCE: EA-ITP-GB, Page 235; EA-AC65-9A, Page 44

5133. ANSWER #4
Note 1 in Figure 28 says to drill the hole to 31/64 of an inch and then to ream to 1/2 inch. The note must be used to answer this question.
REFERENCE: EA-AC65-9A, Page 35

5134. ANSWER #3
In the top view of the clevis pin, the hole is shown with a dimension of .3125 (+ .0005, − .0000). By adding the plus limit to the .3125 dimension, a maximum diameter of 0.3130 is obtained.
REFERENCE: EA-ITP-GB, Page 235

5135. ANSWER #2
In the top view of the clevis pin, the largest cross-section dimension which is given is 7/8 inch. In order to have the clevis pin machined, its starting dimension has to be greater than 7/8 inch. Of the answers given, 1 inch is the minimum diameter that a person could start with.
REFERENCE: EA-ITP-GB, Page 235

5136. ANSWER #3
The bolthole in the clevis pin needs to be drilled to .3125 inch. This decimal is equal to 5/16 as a common fraction.
REFERENCE: EA-ITP-GB, Page 21; EA-AC65-9A, Page 7

5137. ANSWER #4
Sectioning lines, or cross-hatching, is used to show a sectional view of an object. This is done when the object has hidden features that cannot be shown clearly by exterior views.
REFERENCE: EA-ITP-GB, Page 234; EA-AC65-9A, Page 40

5138. ANSWER #1
When a person is trying to find a specific part or dimension on a drawing, like trying to find a road on a road map, zone numbers are used. These numbers, and often letters, are located on the sides of the drawing.
REFERENCE: EA-ITP-GB, Page 237-238; EA-AC65-9A, Page 38

5139. ANSWER #2
Schematics are very helpful in troubleshooting aircraft systems because they show the flow and the relationship of one component part to another. Without an electrical schematic, locating and correcting a fault in an aircraft's electrical system would be almost impossible.
REFERENCE: EA-ITP-GB, Page 230; EA-AC65-9A, Pages 46-49

5140. ANSWER #2
The values identified here are 4.387 inch (the dimension), + .005 (the upper limit), and − .002 (the lower limit). When the upper and lower limits are added together, a value known as the tolerance is obtained. In this case, the tolerance is .007 inch.
REFERENCE: EA-ITP-GB, Page 235

5141. ANSWER #1
Tolerances are the sum of the upper and lower limits. The upper limit given in this question is + .0025 and the lower limit is − .0003. The sum of these limits is .0028.
REFERENCE: EA-ITP-GB, Page 235

5142. ANSWER #2
One of the most helpful features of a schematic is that it does show the flow in a system, be it hydraulic, fuel, oil, or electrical.
REFERENCE: EA-ITP-GB, Page 230; EA-AC65-9A, Page 46-49

5143. ANSWER #3
Print tolerance specifies the standards of manufacturing tolerance to be used for a specific part. It does not take the place of individual tolerances which may be given for certain portions of the part.
REFERENCE: EA-ITP-GB, Pages 236-237

5144. ANSWER #3
The distances which need to be added to get the total distance are as follows: From the top of the plate to the center of the first $^{15}/_{64}$ hole is $^3/_8$ inch. From the center of the first $^{15}/_{64}$ hole to the center of the third $^{15}/_{64}$ hole is $1^3/_4$ inch. From the center of the third $^{15}/_{64}$ hole to the center of the fourth $^{15}/_{64}$ hole is $^1/_8$ inch. From the center of the fourth $^{15}/_{64}$ hole to the bottom of that hole is $^{15}/_{128}$ inch. The total of all these distances is 2.367 inches.
REFERENCE: EA-ITP-GB, Page 235

5145. ANSWER #3
Detail drawings are made as accurately as possible, but the blueprinting process can cause the paper print to shrink and/or stretch. Do not use a blueprint as a pattern. Do a layout using the blueprint dimensions.
REFERENCE: EA-ITP-GB Page 221

5146. ANSWER #1
Exploded-view drawings allow relating of one part to another and make part identification easier, hence their use in parts manuals.
REFERENCE: EA-ITP-GB Page 225

5147. ANSWER #4
An assembly drawing combines parts fabricated from detail drawings into units or sub-assemblies. The units are then installed according to an installation drawing.
REFERENCE: EA-ITP-GB Page 224

5148. ANSWER #3
See explanation number 5147.
REFERENCE: EA-ITP-GB Page224, EA-AC65-9A Page 35

5149. ANSWER #3
Most wiring diagrams will list the sizes of wire that each P/N is fabricated from. In addition, the diagram should list the type of wire and applicability of the finished part.
REFERENCE: EA-ITP-GB Page 228

5150. ANSWER #2
Schematic diagrams, because flow and operations can be illustrated, are used in teaching and troubleshooting.
REFERENCE: EA-ITP-GB Page 230, EA-AC65-9A Page 46

5151. ANSWER #1
To answer this question it is necessary to locate 850 horsepower on the top of the chart. From this value, drop down vertically until the plot line for the 2,000 cubic inch engine is intersected. From the intersection point on this line, extend a line horizontally to the far right side of the chart. On the bottom of the chart, locate the point that represents a BMEP of 185. From this point, extend a line up until it intersects the horizontal line. This will occur at 1800 RPM.
REFERENCE: EA-ITP-GB, Pages 38-40 (general application of charts and graphs)

5152. ANSWER #1
The method of solving this problem is the same as in #5151, except that the final intersection of lines occurs at an RPM plot line. Dropping down from this point, a BMEP of 217 is found.
REFERENCE: EA-ITP-GB, Pages 38-40 (general application of charts and graphs)

5153. ANSWER #2
This probem is solved exactly the same way as #5151. The intersection of the lines will occur at 2100 RPM.
REFERENCE: EA-ITP-GB, Pages 38-40 (general application of charts and graphs)

5154. ANSWER #1
A thorough description of how to use this wire chart is given in the ITP Airframe textbook and in the EA-AC65-15A.
REFERENCE: EA-ITP-AB, Pages 337-338; EA-AC65-15A, Pages 436-439

5155. ANSWER #3
A thorough description of how to use this wire chart is given in the ITP Airframe textbook and in the EA-AC65-15A.
REFERENCE: EA-ITP-AB, Pages 337-338; EA-AC65-15A, Pages 436-439

5156. ANSWER #3
Using Figure 31, find the 500 BHP line and follow it down to the 985 CID. Follow that line scross until it meets the 200 BMEP line coming up from the bottom. These two lines intersect at the answer, 2000RPM.
REFERENCE: Figure 31

5157. ANSWER #1
The chart indicates what tension load a cable should be adjusted to, according to ambient temperature. You should start with the design load on the right and ambient temperature on the bottom. Follow the cable size curve until it crosses the ambient vetical line. Follow this line horizontally to the right to find the cable tension according to ambient temperature.
REFERENCE: Figure 33

5158. ANSWER #1
In this case start with the junction of the 2800 CDI line and the 2000 BHP line. Follow the horizontal line across to the 2200 RPM line. Follow the intersection down the vertical line to find the BMEP.
REFERENCE: Figure 31

5159. ANSWER #1
A thorough description of how to use this wire chart is given in the ITP Airframe textbook and in the EA-AC65-15A.
REFERENCE: EA-ITP-AB, Pages 337-338; EA-AC65-15A, Pages 436-439

5160. ANSWER #1
Find the intermittent curve (3) and the #16 wire size vertical line. Estimate the 25 ampere position on the curve. Follow this intersection to the left and find the answer, 8 feet, in the 28V column.
REFERENCE: Figure 32

5161. ANSWER #2
A thorough description of how to use this wire chart is given in the ITP Airframe textbook and in the EA-AC65-15A.
REFERENCE: EA-ITP-AB, Pages 337-338; EA-AC65-15A, Pages 436-439

5162. ANSWER #1
A thorough description of how to use this wire chart is given in the ITP Airframe textbook and in the EA-AC65-15A.
REFERENCE: EA-ITP-AB, Pages 337-338; EA-AC65-15A, Pages 436-439

5163. ANSWER #3
The use of this chart is described in the ITP Airframe and in the AC65-15A.
REFERENCE: EA-ITP-AB, Pages 282-283; EA-AC65-15A, Page 71

5164. ANSWER #1
To answer this question, proceed as follows: Find on the bottom of the chart (engine speed-RPM) the point that would represent 2300 RPM. From this point, extend a line vertically to the top of the chart. This line will intersect two other needed plot lines, the propeller load specific fuel consumption line and the propeller load horsepower line. Where it intersects the propeller load specific fuel consumption line, extend a line horizontally to the right. This plot will give a value of .47 pounds per hour/horsepower. From where the 2300 RPM line intersects the propeller load horsepower plot, extend a line horizontally to the left. This plot will give a value of 110 horsepower. We now know how many horsepower the engine should be developing and how many pounds per hour of fuel should be consumed per

horsepower. When the two values are multiplied together, a value of 51.7 pounds per hour (fuel consumed) is obtained. Because the question asks for ½ hour of operation, we cut this answer in half and get 25.85 pounds. This is closest to the answer 25.7.
REFERENCE: EA-ITP-P, Page 17

5165. ANSWER #3
This question can be solved in the same way as #5164.
REFERENCE: EA-ITP-P, Page 17

5166. ANSWER #1
Follow the #16 wire size down until it intersects the 10 amp line. Follow this intersection left to the 14V column and find the answer, 10.5 feet. As the #16/10 amp intersection is above the curve, it can be routed in a bundle.
REFERENCE: Figure 32

5167. ANSWER #2
This time, start with the 20 amp line and follow it down to the #12 cable line. Follow this line across to the 28V column and read the answer.
REFERENCE: Figure 32

5168. ANSWER #4
According to FAR 23, the empty weight of an airplane shall include all fixed equipment, permanent ballast, unusable fuel, full oil and full hydraulics.
REFERENCE: EA-ITP-GB, Page 253

5169. ANSWER #1
The useful load of an aicraft is the difference between the maximum weight and the empty weight. It is anything loaded into the airplane above the empty weight.
REFERENCE: EA-ITP-GB, Page 254

5170. ANSWER #1
Answer #3 in this question, the Type Certificate Data Sheet, is a very good source of weight and balance data. The TCD does not give you the empty weight of the aircraft, so to get this information you need to look in the aircraft's weight and balance records.
REFERENCE: EA-ITP-GB, Page 257; EA-AC65-9A, Page 54

5171. ANSWER #1
This distance is the lever arm. The result of the lever arm times the weight is the "moment".
REFERENCE: EA-ITP-GB, PAGE 265; EA-AC65-9A Page 53.

5172. ANSWER #1
The purpose of weighing an aircraft is to determine its empty weight and empty weight center of gravity. When this is being done, the hydraulic reservoirs should be full (FAR 23 requirement).
REFERENCE: EA-ITP-GB, Page 253

5173. ANSWER #1
There is no FAA regulation that requires private aircraft to be weighed periodically. When private aircraft are altered, their new weight and balance is generally calculated mathematically. They are not generally placed on scales and weighed, nor are they required to be.
REFERENCE: EA-AC65-9A, Page 53

5174. ANSWER #4
The Aircraft Specifications of Type Certificate Data Sheets are the documents which identify the equipment needed to maintain the airworthiness certificate.
REFERENCE: EA-ITP-GB, Page 259

5175. ANSWER #4
When an aircraft is type certificated, the aircraft manufacturer designates places on the aircraft where the longitudinal and lateral leveling will be checked. One or both of these leveling means is identified in the Type Certificate Data Sheet.
REFERENCE: EA-ITP-GB, Page 253; EA-AC65-9A, Page 57

5176. ANSWER #2
When an aircraft is weighed, it must be in a flight level attitude. If it is sitting on the scales in a nose high or tail high attitude, the weight shift will be off and the data obtained will not be accurate.
REFERENCE: EA-ITP-GB, Pages 256-257; EA-AC65-9A, Page 61

5177. ANSWER #4
The moment value for an aircraft is obtained by multiplying the weight (pounds) by the arm (distance in inches).
REFERENCE: EA-ITP-GB, Page 254; EA-AC65-9A, Page 54

5178. ANSWER #3
The location of times in an aircraft is given a positive or negative arm value according to their location. Items in front of the datum are at a negative arm and items behind the datum are at a positive arm. The algebraic sign of the item's weight depends on whether the item is being added or taken away from the aircraft. The algebraic sign of the moment is dependent on both the sign of the arm and the sign of the weight.
REFERENCE: EA-ITP-GB, Page 252; EA-AC65-9A, Pages 54-55

5179. ANSWER #3
The difference between the airplane's maximum weight and its empty weight is the useful load.
REFERENCE: EA-ITP-GB, Page 254

5180. ANSWER #3
The farther out on a lever a weight is placed, the greater the moment force (torque). The farther from the center of gravity an object is placed, the greater the effect it will have on the balance of the aircraft.
REFERENCE: EA-ITP-GB, Pages 248-249

5181. ANSWER #4
Regardless of the type of aircraft, weight and balance calculations are done in essentially the same way. The total moment divided by the total weight, in any type of aircraft, will give the center of gravity.
REFERENCE: EA-ITP-GB, Page 268; EA-AC65-9A, Pages 70-71

5182. ANSWER #3
The weighing points used must be indicated on a weighing form because the arm values used in the computation are based on these locations.
REFERENCE: EA-ITP-GB, Page 257; EA-AC65-9A, Pages 57-58

5183. ANSWER #3
All measured arms behind the datum carry a positive (+) value and arms in front of the datum carry a negative (−) value. If the datum is in front of the airplane, there will be no negative arms and the numbers might be easier to work with.
REFERENCE: EA-ITP-GB, Page 252; EA-AC65-9A, Page 54

5184. ANSWER #2
Zero fuel weight is a consideration on aircraft which have a takeoff weight which is greater than the landing weight. When an airplane lands at its primary destination, it should have enough fuel in the tanks to reach an alternate airport and still fly for approximately 30 minutes. If an airplane was allowed to be loaded too heavy with passengers and cargo before fuel was loaded, it might be impossible for the airplane to land at a proper weight and still have enough fuel in the tanks to be legal.
REFERENCE: EA-ITP-GB, Page 254; EA-AC65-9A, Page 58

5185. ANSWER #3
When an airplane is weighed, items such as ground locks, wheel chocks, and jacks are often used to get the airplane properly positioned and secured. If the weight of these items is being absorbed by the scales, it must be mathematically subtracted. The weight of these items is called tare weight.
REFERENCE: EA-ITP-GB, Page 254; EA-AC65-9A, Page 59

5186. ANSWER #3
The maximum weight is the maximum authorized weight of the aircraft and its contents, and is indicated in the specifications.
REFERENCE: EA-ITP-GB, Page 253; EA-AC65-9A, Page 56

5187. ANSWER #3
If the weight and balance of an aircraft is not within limits, there is a very good chance the airplane will crash. The NTSB aircraft accident reports are full of weight and balance caused crashes.
REFERENCE: EA-ITP-GB, Page 246; EA-AC65-9A, Page 53

5188. ANSWER #3
39.00″ is 1.63″ aft of the original C.G.
REFERENCE: EA-ITP-GB, Page 268; EA-AC65-9A, Pages 66-67

5189. ANSWER #1
Until March 1, 1978 empty weight included only residual oil. Since March of '78, full oil has been included in the empty weight of small airplanes.
REFERENCE: EA-ITP-GB, Page 253; EA-FAR 23.29

5190. ANSWER #1
Trimming of a helicopter which has less than ideal balance is done with the cyclic pitch. If the balance problem is extreme (exceeding the limits), the effectiveness of the cyclic pitch will be reduced or possibly lost.
REFERENCE: EA-AC65-9A, Page 71

5191. ANSWER #1
The difference between an aircraft's empty weight and its maximum weight is the amount of weight the aircraft can carry of useful items. This is known as the useful load.
REFERENCE: EA-ITP-GB, Page 254; EA-AC65-9A, Page 56

5192. ANSWER #3
Although the Type Certificate Data Sheet on an aircraft does not give its empty weight or useful load, it does list the aircraft's maximum weight.
REFERENCE: EA-ITP-GB, Page 258; EA-AC65-9A, Page 56

5193. ANSWER #3
The answer is computed utilizing the standard weight and balance formulas.
REFERENCE: EA-ITP-GB, Page 268; EA-AC65-9A, Pages 66-67

5194. ANSWER #4
While the actual weight and balance computations and methodology are the same for helicopters, shifts in balance are much more critical, hence a shorter CG range.
REFERENCE: EA-ITP-GB Page 268

5195. ANSWER #2
Unusable fuel is included in the empty weight of an aircraft. It is assumed that this question is talking about empty weight.
REFERENCE: EA-ITP-GB, Page 253; EA-AC65-9A, Page 56

5196. ANSWER #4
The answer is computed utilizing the standard weight and balance formulas.
REFERENCE: EA-ITP-GB, Pages 264-265; EA-AC65-9A, Page 67

5197. ANSWER #4
The answer is computed utilizing the standard weight and balance formulas.
REFERENCE: EA-ITP-GB, Page 268; EA-AC65-9A, Pages 66-67

5198. ANSWER #2
For the center of gravity of the airplane in question not to change, the moments forward and aft of the C.G. must balance out. The two boxes added aft of the C.G. would have moments of 40 and 10 (10 pounds × 4 feet and 5 pounds × 2 feet). This is a total moment of 50. The box loaded forward must also have a moment of 50. The 20 pound box would need a distance of 2.5 feet to have a moment of 50.
REFERENCE: EA-ITP-GB, Pages 248-249

5199. ANSWER #1
The reason an aircraft might have (most do not) an empty weight center of gravity range is to indicate whether adverse check calculations are needed to check the loading of the airplane.
REFERENCE: EA-ITP-GB, Page 253; EA-AC65-9A, Page 57

5200. ANSWER #3
The answer is computed utilizing the standard weight and balance formulas.
REFERENCE: EA-ITP-GB, Page 268; EA-AC65-9A, Pages 66-67

5201. ANSWER #1
The answer is computed utilizing the standard weight and balance formulas.
REFERENCE: EA-ITP-GB, Page 249; EA-AC65-9A, Page 54

5202. ANSWER #3
The answer is computed utilizing the standard weight and balance formulas.
REFERENCE: EA-ITP-GB, Page 249; EA-AC65-9A. Page 63

5203. ANSWER #2
The answer is computed utilizing the standard weight and balance formulas.
REFERENCE: EA-ITP-GB, Pages 267-268

5204. ANSWER #2
The answer is computed utilizing the standard weight and balance formulas.
REFERENCE: EA-ITP-GB, Pages 261-262; EA-AC65-9A, Pages 63-64

5205. ANSWER #4
When a rearward extreme condition check is performed, items of useful load in front of the rearward limit should be loaded to the minimum and items of useful load behind the rearward limit should be loaded to the maximum.
REFERENCE: EA-ITP-GB, Page 267; EA-AC65-9A, Page 66

5206. ANSWER #1
The answer is computed utilizing the standard weight and balance formulas.
REFERENCE: EA-ITP-GB, Pages 261-262; EA-AC65-9A, Pages 63-64

5207. ANSWER #4
The answer is computed utilizing the standard weight and balance formulas.
REFERENCE: EA-ITP-GB, Page 268; EA-AC65-9A, Pages 66-67

5208. ANSWER #4
The answer is computed utilizing the standard weight and balance formulas.
REFERENCE: EA-ITP-GB, Page 268; EA-AC65-9A, Pages 66-67

5209. ANSWER #2
When an adverse forward weight and balance check is done, maximum weight of useful load items is used forward of the forward C.G. limit and minimum weight of useful load items is used aft of the forward C.G. limit.
REFERENCE: EA-ITP-GB, Page 266; EA-AC65-9A, Page 66

5210. ANSWER #1
The answer is computed utilizing the standard weight and balance formulas.
REFERENCE: EA-ITP-GB, Page 268; EA-AC65-9A, Pages 66-67

5211. ANSWER #4
The answer is computed utilizing the standard weight and balance formulas.
REFERENCE: EA-ITP-GB, Pages 261-262; EA-AC65-9A, Pages 63-64

5212. ANSWER #2
The final dash number in these AN-818 coupling nuts indicates the size tube it will fit. The numbers are in $1/16$ inch increments, so a dash 8 would be $1/2$ inch.
REFERENCE: EA-ITP-GB, Page 276; EA-AC65-9A, Page 105

5213. ANSWER #1
Hydraulic lines left unsupported can easily come in contact with sharp or damaging objects, leaving them damaged and subject to failure.
REFERENCE: EA-ITP-GB, Page 295; EA-AC65-9A, Page 115

5214. ANSWER #3
The steps identified in this question are given word for word in the EA-AC65-9A.
REFERENCE: EA-ITP-GB, Pages 281-282; EA-AC65-9A, Page 111

5215. ANSWER #1
A severely damaged line should be replaced; however, it may be repaired by cutting out the damaged section and inserting a tube section of the same size and material. Flare both ends of the undamaged and replacement tube sections and make the connection by using standard unions, sleeves, and tube nuts.
REFERENCE: EA-ITP-GB, Pages 278-281; EA-AC43.13-1A, Page 166(c)

5216. ANSWER #4
The double flare is smoother and more concentric than the single flare and therefore seals better. It is also more resistant to the shearing effect of torque.
REFERENCE: EA-ITP-GB, Page 282; EA-AC65-9A, Page 111

5217. ANSWER #4
When failure occurs in a flexible hose equipped with swaged end fittings, the entire assembly must be replaced. Obtain a new hose assembly of the correct size and length, complete with factory-installed end fittings.
REFERENCE: EA-ITP-GB, Page 291; EA-AC65-9A, Page 114

5218. ANSWER #1
AN fittings have a shoulder between the end of the threads and the flare cone. The AC fitting does not have this shoulder.
REFERENCE: EA-ITP-GB, Page 289; EA-AC65-9A, Page 104

5219. ANSWER #4
Flexible fluid lines must have between five and eight percent slack to allow for the change in dimensions of the hose caused by the pressure and for vibration and expansion of the airframe.
REFERENCE: EA-ITP-GB, Page 295; EA-AC65-9A, Page 115

5220. ANSWER #4
When a flexible hose is installed, it should have five to eight percent slack. For a 50 inch span, a hose would need to be $52^{1/2}$ inches long.
REFERENCE: EA-ITP-GB, Page 295; EA-AC65-9A, Page 115

5221. ANSWER #2
A flexible hose must be installed with slack to allow for vibration and movement as the hose is pressurized.
REFERENCE: EA-ITP-GB, Page 295; EA-AC65-9A, Page 115

5222. ANSWER #4
Flared-tube fittings, for identification purposes, are color-coded. Steel fittings are black, and aluminum are blue.
REFERENCE: EA-ITP-GB, Page 289; EA-AC65-9A, Page 107

5223. ANSWER #4
Soft aluminum tubing (1100, 3003, or 5052) under 1/4 inch outside diameter may be bent by hand.
REFERENCE: EA-ITP-GB, Page 278; EA-AC43.13-1A, Page 165(a)

5224. ANSWER #1
The 0.072 identification is the wall thickness of the tube. To find the inside diameter of the tube, two wall thicknesses must be subtracted from the outside diameter.
REFERENCE: EA-ITP-GB, Page 276

5225. ANSWER #3
The AN817 single-piece nut is not recommended because it tends to wipe or iron the flare as it is being tightened. The AN818 nut with the sleeve is the most popular type.
REFERENCE: EA-ITP-GB, Page 281

5226. ANSWER #2
Final tightening of a MS flareless fitting depends upon the tubing. For aluminum alloy tubing up to and including 1/2 inch outside diameter, the nut is tightened from one to one-and-one-sixth turns.
REFERENCE: EA-AC65-9A, Page 112

5227. ANSWER #2
Automotive fittings have a 45° angle of flare. To prevent the use of these fittings, aircraft fittings have a 37° angle of flare.
REFERENCE: EA-ITP-GB, Page 280; EA-AC65-9A, Page 111

5228. ANSWER #4
Scratches or nicks no deeper than 10% of the wall thickness in aluminum alloy tubing, that are not in the heel of a bend, may be repaired by burnishing with hand tools.
REFERENCE: EA-ITP-GB, Page 298; EA-AC65-9A, Page 166(c)

5229. ANSWER #4
When there is relative motion between the components that are joined by a fluid line, a flexible hose must be used.
REFERENCE: EA-ITP-GB, Page 284; EA-AC65-9A, Page 100

5230. ANSWER #2
If it is impossible to separate fluid lines from electrical wire bundles, the wire bundle must be routed above the fluid line, and it must be clamped securely to the structure to prevent it from contacting the fluid line.
REFERENCE: EA-ITP-GB, Page 295; EA-AC43.13-1A, Page 205(f)

5231. ANSWER #3
It is extremely important that a nut never be used to pull the tube up against the fitting, as vibration or expansion of the aircraft can cause the tube to fail at the flare.
REFERENCE: EA-ITP-GB, Page 280

5232. ANSWER #4
The size of a rigid fluid line is measured by its outside diameter, but the size of a flexible line is determined by its inside diameter.
REFERENCE: EA-ITP-GB, Page 285; EA-AC65-9A, Page 102

5233. ANSWER #2
Scratches or nicks no deeper than 10% of the wall thickness in aluminum tubing, that are not in the heel of a bend, may be repaired by burnishing with hand tools.
REFERENCE: EA-ITP-GB, Page 298; EA-AC43.13-1A, Page 166(c)

5234. ANSWER #2
Operating strengths include wide temperature range, compatibility with fluids and gasses used in aviation, little resistance to flow, and nearly limitless shelf life.
REFERENCE: EA-ITP-GB, Page 287; EA-AC65-9A, Page 100

5235. ANSWER #1
The 2024-T and 5052-0 aluminum alloy materials are used in general purpose systems of low and medium pressures, such as hydraulic and pneumatic 1,000 to 1,500 PSI systems and fuel and oil lines.
REFERENCE: EA-ITP-GB, Page 276; EA-AC65-9A, Page 100

5236. ANSWER #3
Corrosion-resistant steel tubing, either annealed or 1/4 hard, is used extensively in high pressure hydraulic systems for the operation of landing gear, flaps, brakes, and the like.
REFERENCE: EA-ITP-GB, Page 276; EA-AC65-9A, Page 119

5237. ANSWER #4
When using bonded clamps to secure metal hydraulic, fuel, and oil lines, any paint or anodizing should be removed from the portion of the tube where the clamp will be located.
REFERENCE: EA-AC65-9A, Page 119

5238. ANSWER #2
All tubing installations should have a least one bend between the fittings to absorb vibrations and the stress of the line under pressure.
REFERENCE: EA-ITP-GB, Page 295; EA-AC65-9A, Pages 115-116

5239. ANSWER #3
Lines which carry materials that are physically dangerous may be marked with the word (abbreviation) PHDAN.
REFERENCE: EA-ITP-GB, Page 298; EA-AC65-9A, Page 103

5240. ANSWER #1
Teflon® hose is unaffected by any known fuel, petroleum, or synthetic base oils, alcohol, coolants, or solvents commonly used in aircraft. Its principal advantage, however, is its operating strength.
REFERENCE: EA-ITP-GB, Page 287; EA-AC65-9A, Page 102

5241. ANSWER #2
Bernoulli's principle states that when a fluid is in motion with a constant energy input, as the pressure energy increases, the velocity energy decreases.
REFERENCE: EA-ITP-GB, Page 82; EA-AC65-9A, Page 242

5242. ANSWER #3
Bonded clamps are used to secure metal hydraulic, fuel, and oil lines in place. Unbonded clamps should be used only for securing wiring.
REFERENCE: EA-AC65-9A, Page 119

5243. ANSWER #3
The use of any type of hose will depend on the manufacturer's specifications.
REFERENCE: EA-AC65-9A, Page 99

5244. ANSWER #3
One of the most important methods of non-destructive inspection available to the mechanic is radiographic inspection. It allows the technician to see the inside of a structure that is opaque to light.
REFERENCE: EA-ITP-GB, Page 442; EA-AC65-9A, Page 479

5245. ANSWER #4
In the magnetic particle method of nondestructive test, flaws can be detected both on and below the surface. It cannot be used on nonmagnetic materials.
REFERENCE: EA-ITP-GB, Page 429; EA-AC65-9A, Page 469

5246. ANSWER #4
Dye penetrant works by flowing into a crack or flaw which might be so small that it is not visible to the naked eye. The defect or fault, however, must be on the surface.
REFERENCE: EA-ITP-GB, Page 425; EA-AC65-9A, Page 477

5247. ANSWER #4
The eddy current inspection works on the principle of determining the ease with which a material will accept induced current. An object which has intergrannular corrosion will not accept the current very well, and this shows up on the meter as a varying needle deflection.
REFERENCE: EA-ITP-GB, Page 437; EA-AC65-9A, Page 485

5248. ANSWER #1
As was stated in the explanation to question #5244, the radiographic inspection method is very good for checking internal condition.
REFERENCE: EA-ITP-GB, Page 442; EA-AC65-9A, Page 479

5249. ANSWER #3
Refer back to questions 5245 and 5246.
REFERENCE: EA-ITP-GB, Pages 425 and 429; EA-AC65-9A, Pages 469 and 477

5250. ANSWER #2
With the radiographic inspection method, you are able to look inside a structure. This means that little disassembly is required.
REFERENCE: EA-ITP-GB, Page 442; EA-AC65-9A, Page 479

5251. ANSWER #3
The continuous method of magnetic particle inspection will reveal more nonsignificant discontinuities than the residual procedure, so it is used most often.
REFERENCE: EA-ITP-GB, Pages 436-437; EA-AC65-9A, Page 472

5252. ANSWER #2
Before attempting to use an x-ray machine, the operator must know the unit's capabilities and know how to operate it.
REFERENCE: EA-AC65-9A, Page 479

5253. ANSWER #2
The continuous method may be used in practically all circular and longitudinal magnetization procedures. The continuous procedure provides greater sensitivity than the residual procedure.
REFERENCE: EA-AC65-9A, Page 472

5254. ANSWER #4
When the magnetic particle inspection method is used, there may be no flux leakage if the discontinuity is very far below the surface. This means there would be no indication on the surface.
REFERENCE: EA-AC65-9A, Page 469

5255. ANSWER #4
Cracks, splits, bursts, tears, seams, voids, and pipes are all formed by an actual parting or rupture of the solid metal.
REFERENCE: EA-AC65-9A, Page 469

5256. ANSWER #3
In general, the residual procedure is used only with steels which have been heat treated for stressed applications.
REFERENCE: EA-AC65-9A, Page 472

5257. ANSWER #2
The various types of indicating mediums available for magnetic particle inspection may be divided into two general types: wet process materials and dry process materials.
REFERENCE: EA-ITP-GB, Page 436; EA-AC65-9A, Page 476

5258. ANSWER #3
Magnetic particle inspection is a method of detecting invisible cracks and other defects in ferro-magnetic materials, such as iron and steel.
REFERENCE: EA-ITP-GB, Page 429; EA-AC65-9A, Page 469

5259. ANSWER #3
A part can be demagnetized using AC or DC. The procedure described in answer #3 is the best of the choices.
REFERENCE: EA-ITP-GB, Page 437; EA-AC65-9A, Page 477

5260. ANSWER #1
Circular magnetization will locate defects running approximately parallel to the axis of the part. Longitudinal magnetization will locate defects running approximately 90° to the axis of the part. If the defect is running at a 45° angle, either method will detect it.
REFERENCE: EA-ITP-GB, Page 432; EA-AC65-9A, Page 471

5261. ANSWER #1
Dye penetrants need surface cracks in order to detect the flaw. Dye penetrants work very well with aluminum castings and forgings.
REFERENCE: EA-ITP-GB, Page 425; EA-AC65-9A, Page 477

5262. ANSWER #3
The smaller the crack in an object, the longer the penetrant needs to be in contact with the flaw. For this method of inspection to work, the penetrant has to flow into the crack.
REFERENCE: EA-ITP-GB, Page 427; EA-AC65-9A, Page 478

5263. ANSWER #1
Circular magnetization will reveal defects running approximately parallel to the axis of the part. Longitudinal magnetization will reveal defects running approximately 90° to the axis of the part. To reveal all possible defects, it is necessary to magnetize with both methods.
REFERENCE: EA-ITP-GB, Page 432; EA-AC65-9A, Page 471

5264. ANSWER #1
When using the dye penetrant method of inspecting, the defect must be on the surface. The penetrant must have an opening to flow into.
REFERENCE: EA-ITP-GB, Page 425; EA-AC65-9A, Page 477

5265. ANSWER #4
The penetrant will begin to bleed out of any fault as soon as the surface penetrant has been removed, so to pinpoint the location of the fault, the surface to be inspected must be covered with a developer as soon as possible.
REFERENCE: EA-ITP-GB, Page 427; EA-AC65-9A, Page 478

5266. ANSWER #1
Dye penetrant inspection will only detect surface defects.
REFERENCE: EA-ITP-GB, Page 425; EA-AC65-9A, Page 477

5267. ANSWER #1
Only when the surface being inspected is perfectly clean can the penetrant be assured of getting into any cracks or faults.
REFERENCE: EA-ITP-GB, Page 426; EA-AC65-9A, Page 478

5268. ANSWER #4
The success and reliability of a penetrant inspection depends upon the thoroughness with which the part was prepared. Improper cleaning is the primary cause of bad results.
REFERENCE: EA-ITP-GB, Pages 425-426; EA-AC65-9A, Page 478

5269. ANSWER #1
By using decreasing alternating current on a part which needs to be demagnetized, the magnetic poles are constantly changing because of the AC, and the strength of the fields is decreasing because of the decreasing AC. This effectively demagnetizes the part.
REFERENCE: EA-ITP-GB, Page 437; EA-AC65-9A, Page 477

5270. ANSWER #1
The smaller the defect, the longer the penetrating time. Fine crack-like apertures require a longer penetrating time than defects such as pores.
REFERENCE: EA-ITP-GB, Page 427; EA-AC65-9A, Page 478

5271. ANSWER #2
The results obtained by heat treatment depend to a great extent on the structure of the metal and on the manner in which the structure changes when the metal is heated and cooled. A pure metal cannot be hardened by heat treatment because there is little change in its structure when heated.
REFERENCE: EA-ITP-GB, Page 380; EA-AC65-9A, Page 205

5272. ANSWER #4
Indications of subsurface inclusions are usually broad and fuzzy. Close examination generally reveals their lack of definition is from several parallel lines rather than a single line.
REFERENCE: EA-AC65-9A, Page 474

5273. ANSWER #2
Petroleum base solvents, usually in the form of spray cans, are used to clean a part before the application of penetrants.
REFERENCE: EA-ITP-GB, Pages 426-427; EA-AC65-9A, Page 478

5274. ANSWER #3
Fatigue cracks give sharp, clear patterns, generally uniform and unbroken throughout their length and with good buildup.
REFERENCE: EA-AC65-9A, Page 473

5275. ANSWER #3
Fatigue cracks are found in parts that have been in service but are never found in new parts. They are usually in highly stressed areas of the part.
REFERENCE: EA-AC65-9A, Page 473

5276. ANSWER #2
When dye penetrant inspection is used, the defect must be open to the surface.
REFERENCE: EA-ITP-GB, Page 425; EA-AC65-9A, Page 477

5277. ANSWER #1
Longitudinal magnetization will locate defects running approximately 90° (perpendicular) to the axis of the part.
REFERENCE: EA-ITP-GB, Page 432; EA-AC65-9A, Page 471

5278. ANSWER #1
Circular magnetization will locate defects running approximately parallel to the axis of the part.
REFERENCE: EA-ITP-GB, Page 432; EA-AC65-9A, Page 471

5279. ANSWER #2
Intergranular corrosion attacks aluminum alloy which has been improperly heat-treated. Waiting too long to quench the aluminum alloy is the biggest cause of this corrosion.
REFERENCE: EA-ITP-GB, Page 405; EA-AC65-9A, Page 178

5280. ANSWER #4
Aluminum alloy must be immediately quenched in water to solidify the alloy elements into extremely tiny grains. If this is not done, the corrosion resistance will be impaired.
REFERENCE: EA-ITP-GB, Page 405; EA-AC65-9A, Pages 213-214

5281. ANSWER #1
Casehardening is ideal for parts which require a wear-resistant surface and, at the same time, must be tough enough internally to withstand the applied loads.
REFERENCE: EA-ITP-GB, Page 389; EA-AC65-9A, Page 211

5282. ANSWER #3
Casehardening is used to form a corrosion-resistant outer surface while allowing the material to remain soft and tough internally.
REFERENCE: EA-ITP-GB, Page 389; EA-AC65-9A, Page 211

5283. ANSWER #3
When not quenched fast enough, the dissimilarity between the aluminum and the alloying elements can cause intergranular corrosion to form.
REFERENCE: EA-ITP-GB, Page 405; EA-AC65-9A, Pages 213 and 214

5284. ANSWER #4
Aluminum alloy must be immediately quenched in water to solidify the alloy elements into extremely tiny grains. If this is not done, the corrosion resistance will be impaired.
REFERENCE: EA-ITP-GB, Page 405; EA-AC65-9A, Pages 213-214

5285. ANSWER #1
Nitriding is a form of casehardening. Unlike other casehardening processes, before nitriding, the part is heat treated.
REFERENCE: EA-ITP-GB, Page 391; EA-AC65-9A, Page 212

5286. ANSWER #2
The proper application of temperature, and the proper quenching medium will cause variations in the carbon content of the steel, resulting in varying tempers and tensile strengths.
REFERENCE: EA-ITP-GB, Pages 387 and 388; EA-AC65-9A, Pages 208-210

5287. ANSWER #4
Annealing of steel is accomplished by heating the metal to just above the upper critical point, soaking at that temperature, and cooling it very slowly in the furnace. Normalizing is accomplished by heating the steel above the upper critical point and cooling in still air.
REFERENCE: EA-ITP-GB, Page 388; EA-AC65-9A, Page 211

5288. ANSWER #4
When steel is hardened by quenching, it is usually too hard and brittle for the use for which it is intended, and the rapid cooling usually causes stresses within the steel that must be relieved. To relieve the stresses and reduce the brittleness, the steel is tempered.
REFERENCE: EA-ITP-GB, Page 388; EA-AC65-9A, Page 211

5289. ANSWER #4
Because corrosion is not generally as critical in large forgings as it is for smaller or thinner sections, hot water quenching may be used to help alleviate cracking and distortion.
REFERENCE: EA-ITP-GB, Page 381; EA-AC65-9A, Page 214

5290. ANSWER #2
Some aluminum alloys, such as 2017 and 2024, develop their full properties as a result of solution heat treatment followed by about 4 days of natural aging. Other alloys, such as 7075, require artificial aging.
REFERENCE: EA-AC65-9A, Pages 212-213

5291. ANSWER #3
2024 rivets are purchased from the manufacturer in the heat-treated condition. Because they become hardened in ten minutes after quenching, they must be re-heat-treated before being driven.
REFERENCE: EA-AC65-9A, Page 216

5292. ANSWER #3
Precipitation heat treatment of magnesium improves the yield strength, the hardness, and the corrosion resistance.
REFERENCE: EA-AC65-9A, Page 217

5293. ANSWER #2
The minimum tempering time at temperature is one hour, up to one inch of thickness, with an additional hour for each inch of thickness.
REFERENCE: EA-AC65-9A, Page 211

5294. ANSWER #1
The steels best suited to casehardening are the low-carbon and low-alloy steels. If high-carbon steel is casehardened, the hardness penetrates the core and causes brittleness.
REFERENCE: EA-ITP-GB, Page 385; EA-AC65-9A, Page 211

5295. ANSWER #4
Depending on the thickness of the cladding, clad aluminum sheets can only be reheated one to three times, while unclad aluminum may be repeatedly reheated.
REFERENCE: EA-AC65-9A, Page 214

5296. ANSWER #2
The two ground surfaces should be parallel to each other and, when installed on the tester, perpendicular to the axis of penetration.
REFERENCE: EA-AC65-9A, Page 220

5297. ANSWER #1
At ordinary temperatures, the carbon in steel exists in the form of particles of iron carbide scattered throughout an iron matrix known as "ferrite". The number, size, and distribution of these particles determine the hardness of the steel.
REFERENCE: EA-ITP-GB, Page 387-388; EA-AC65-9A, Page 210

5298. ANSWER #1
The general rule states that nuts and bolts, except corrosion-resistant steel, should not be lubricated when referencing their torque value.
REFERENCE: EA-ITP-GB, Page 366; EA-AC65-9A, Page 135

5299. ANSWER #2
Because of its fireproof qualities, stainless steel is used for the construction of engine firewalls.
REFERENCE: EA-AC65-9A, Page 196

5300. ANSWER #2
Whenever possible, the bolt should be placed with the head on top or in the forward position. This positioning tends to prevent the bolt from slipping out if the nut is accidently lost.
REFERENCE: EA-AC65-9A, Page 132

5301. ANSWER #2
The terms "Alclad and Pureclad" are used to designate sheets that consist of an aluminum alloy core coated with a layer of pure aluminum to a depth of approximately $5^1/2$% on each side.
REFERENCE: EA-AC65-9A, Page 201

5302. ANSWER #3
Self-locking nuts should not be used in any location where the nut or the bolt is subject to rotation.
REFERENCE: EA-ITP-GB, Page 318; EA-AC65-9A, Page 127

5303. ANSWER #4
Steel containing carbon in percentages ranging from 0.10 to 0.30 percent is classed as low-carbon steel. The equivalent numbers range from 1010 to 1030.
REFERENCE: EA-ITP-GB, Page 387; EA-AC65-9A, Page 196

5304. ANSWER #3
A corrosion-resistant AN standard steel bolt is identified with a dash on its head.
REFERENCE: EA-ITP-GB, Page 313; EA-AC65-9A, Page 122

5305. ANSWER #3
In general, bolt grip lengths should equal the material thickness. However, bolts of slightly greater grip length may be used provided washers are placed under the nut or the bolthead.
REFERENCE: EA-AC43.13-1A, Page 115(b)

5306. ANSWER #3
See explanation in #5305.
REFERENCE: EA-AC43.13-1A, Page 115(b)

5307. ANSWER #2
The total thickness of the two plates is $1^1/4$ inches. The grip length of the bolt should be the same.
REFERENCE: EA-AC43.13-1A, Page 115(b)

5308. ANSWER #4
Figure 5.2 in the EA-AC43.13-1A provides a recommended torque value table.
REFERENCE: EA-AC43.13-1A, Page 118

5309. ANSWER #3
A clevis bolt is used when there are only shear loadings.
REFERENCE: EA-ITP-GB, Page 314

5310. ANSWER #1
This is a corrosion-resistant AN steel bolt. Its head design is shown in view "C" of Figure 359.
REFERENCE: EA-ITP-GB, Page 313; EA-AC65-9A, Page 122

5311. ANSWER #2
Because the clevis bolt is not intended for tension loads, the shear nut is only tightened to a snug fit, and then safetied with a cotter pin.
REFERENCE: EA-ITP-GB, Pages 314-315

5312. ANSWER #4
A cross inside a triangle indicates a close tolerance bolt.
REFERENCE: EA-ITP-GB, Page 314; EA-AC65-9A, Page 122

5313. ANSWER #4
Clevis bolts are used only for shear load applications.
REFERENCE: EA-ITP-GB, Pages 314-315

5314. ANSWER #3
The NAS close tolerance bolt has a triangle on its head with an "X" inside it.
REFERENCE: EA-ITP-GB, Page 314; EA-AC65-9A, Page 122

5315. ANSWER #2
The terms "Alclad and Pureclad" are used to designate sheets that consist of an aluminum alloy core coated with a layer of pure aluminum to a depth of approximately 5½% on each side.
REFERENCE: EA-AC65-9A, Page 201

5316. ANSWER #4
Aluminum with a code #1100 is said to be pure aluminum.
REFERENCE: EA-ITP-GB, Page 380; EA-AC65-9A, Page 200

5317. ANSWER #3
A Class 3 fit is a medium fit, and it is the class in which most aircraft bolts are manufactured.
REFERENCE: EA-ITP-GB, Page 380; EA-AC65-9A, Page 121

5318. ANSWER #4
Two thousand-series aluminum alloys use copper as the major alloying element.
REFERENCE: EA-ITP-GB, Page 380; EA-AC65-9A, Page 199

5319. ANSWER #2
A fiber lock nut uses a fiber insert which is not threaded until the first bolt passes through it.
REFERENCE: EA-ITP-GB, Page 318; EA-AC65-9A, Pages 128-129

5320. ANSWER #3
The puddle has a tendency to boil during the welding operation if an excessive amount of acetylene is used. This often leaves slight bumps along the center and craters at the finish of the weld.
REFERENCE: EA-AC65-9A, Page 487

5321. ANSWER #3
A weld which has had insufficient heat (cold weld) will appear rough and irregular and its edges will not be feathered into the base metal.
REFERENCE: EA-AC65-9A, Page 487

5322. ANSWER #2
When rewelding a joint, the old welded joint should be cut out and replaced with one that is properly gusseted and reinforced.
REFERENCE: EA-AC43.13-1A, Page 47

5323. ANSWER #1
Normalizing removes the internal stresses set up by heat treating, welding, casting, forming, or machining.
REFERENCE: EA-ITP-GB, Page 388; EA-AC65-9A, Page 211

5324. ANSWER #3
When checking the condition of a completed weld, there should be no signs of blowholes, porosity, or projecting globules. If there is, the weld should be removed and rewelded.
REFERENCE: EA-AC43.13-1A, Page 30

5325. ANSWER #3
A good bead with proper penetration and good fusion will be straight across the sheet and will have a smoothly crowned surface that flairs evenly into the base metal.
REFERENCE: EA-ITP-AB, Page 197; EA-AC43.13-1A, Page 30

5326. ANSWER #1
A weld which shows projecting globules is not acceptable.
REFERENCE: EA-AC43.13-1A, Page 30

5327. ANSWER #2
For a weld to be sound, the metals must flow together and become one. According to the EA-AC43.13-1A, the depth of penetration of the weld must ensure that fusion of the base metal and the filler rod takes place.
REFERENCE: EA-AC43.13-1A, Page 30

5328. ANSWER #1
According to the EA-AC43.13-1A, a complete weld should have no oxide formed on the base metal at a distance of more than ½ inch from the weld.
REFERENCE: EA-AC43.13-1A, Page 30

5329. ANSWER #2
A butt joint weld is made by placing two pieces of material edge to edge, so that there is no overlapping, and then welded.
REFERENCE: EA-ITP-AB, Page 197; EA-AC65-15A, Pages 255-256

5330. ANSWER #2
A double butt welded joint is one where a bead has been applied on both sides of the joint.
REFERENCE: EA-ITP-AB, Page 197; EA-AC65-15A, Pages 255-256

5331. ANSWER #1
The lap joint, either single or double, is one where the two pieces of material are overlapped and welded at the joint (single lap) or joints (double lap).
REFERENCE: EA-ITP-AB, Page 198; EA-AC65-15A, Pages 256-257

5332. ANSWER #2
End play on an axle is measured by having a dial indicator rod in contact with the axle (under a preload) and then rotating the axle. End play will cause the dial indicator rod to move in or out, which will be read on the face of the dial indicator as thousandths or possibly ten-thousandths of an inch end play.
REFERENCE: EA-AC65-12A, Page 420

5333. ANSWER #2
The reading in this question is coming from a Vernier micrometer. The correct way of taking the reading is explained in the EA-AC65-9A and in the ITP General Textbook.
REFERENCE: EA-ITP-GB, Page 372; EA-AC65-9A, Page 545-546

5334. ANSWER #2
The proper way to read a Vernier caliper is explained in the ITP General Textbook and in the EA-AC65-9A.
REFERENCE: EA-ITP-GB, Page 374; EA-AC65-9A, Page 546

5335. ANSWER #3
A thickness gage, often called a feeler gage, is used to check the clearance between surface plates and the object being checked.
REFERENCE: EA-ITP-P, Page 71; EA-AC65-12A, Page 429

5336. ANSWER #4
The threads in a micrometer caliper are ground to extreme accuracy and have a pitch of 40 threads to the inch. One revolution of the thimble advances the spindle twenty-five thousandths of an inch.
REFERENCE: EA-ITP-GB, Page 370; EA-AC65-9A, Page 546

5337. ANSWER #3
The graduations on a Vernier micrometer are one ten-thousandth of an inch (.0001).
REFERENCE: EA-ITP-GB, Page 372; EA-AC65-9A, Page 546

5338. ANSWER #1
The reading shown in this question is on a Vernier micrometer. The proper way to read a Vernier micrometer is explained in the ITP General Textbook and in the EA-AC65-9A.
REFERENCE: EA-ITP-GB, Page 372; EA-AC65-9A, Page 546

5339. ANSWER #2
The reading shown in this question is on a Vernier micrometer. The proper way to read a Vernier micrometer is explained in the ITP General Textbook and in the EA-AC65-9A.
REFERENCE: EA-ITP-GB, Page 372; EA-AC65-9A, Page 546

5340. ANSWER #1
The center head on the combination set is used to find the center of shafts or other cylindrical work.
REFERENCE: EA-ITP-GB, Page 368; EA-AC65-9A, Page 543

5341. ANSWER #4
Telescoping gages and micrometers are generally used to measure the diameter of a hole. In this case, however, the hole is too small to use a telescoping gage, so a ball gage needs to be used.
REFERENCE: EA-AC65-12A, Page 420

5342. ANSWER #3
The reading shown in this question is on a Vernier micrometer. The proper way to read a Vernier micrometer is explained in the ITP General Textbook and in the EA-AC65-9A.
REFERENCE: EA-ITP-GB, Page 372; EA-AC65-9A, Page 546

5343. ANSWER #1
A machinist scale is used to set the dimension on a divider.
REFERENCE: EA-ITP-GB, Page 368; EA-AC65-9A, Page 544

5344. ANSWER #1
Special standard gages are used to check the zero adjustment on micrometers.
REFERENCE: EA-ITP-GB, Page 372

5345. ANSWER #3
By taking a dimensional check at various locations on a crankpin or main bearing journal, it is possible to determine the degree of out-of-round.
REFERENCE: EA-ITP-P, Page 65

5346. ANSWER #2
Side clearance of piston rings is checked with a thickness gage, often called a feeler gage.
REFERENCE: EA-ITP-P, Page 70

5347. ANSWER #2
Runout on a crankshaft is checked with a dial indicator (dial gage).
REFERENCE: EA-AC65-12A, Page 428

5348. ANSWER #4
A telescopic gage (T-gage) is placed in the rocker arm bearing and allowed to expand. It is locked in the expanded position, and this dimension is read with a micrometer.
REFERENCE: EA-AC65-12A, Page 420

5349. ANSWER #3
With the arbors installed, the rod is layed across the parallel blocks on a surface plate and a check is made to see if a feeler gage can be passed between the arbor and the block.
REFERENCE: EA-ITP-P, Page 71

5350. ANSWER #2
This check is known as a side clearance check, and it is done with a thickness gage (feeler gage).
REFERENCE: EA-ITP-P, Page 70

5351. ANSWER #2
Piston ring gap is measured with a thickness gage (feeler gage).
REFERENCE: EA-ITP-P, Page 70

5352. ANSWER #2
A valve stretch gage is laid against the stem of the valve, with a curved section of the gage following the curvature of valve face. If there is a gap between the gage and the face of the valve, the valve has stretched.
REFERENCE: EA-AC65-12A, Page 418

5353. ANSWER #2
By taking a dimensional check at various locations (different diameter positions) on a piston pin, it is possible to check it for out-of-round wear. This is done with a micrometer.
REFERENCE: EA-AC65-12A, Pages 418-420

5354. ANSWER #4
A hung start is one in which the engine starts but does not accelerate enough for the compressor to supply sufficient air for the engine to become self-accelerating. If a hung start occurs, the engine should be shut down.
REFERENCE: EA-ITP-GB, Page 472; EA-AC65-9A, Page 494

5355. ANSWER #3
The person in charge of the towing operation should verify that, on aircraft with a steerable nosewheel, the locking scissors are set to full swivel for towing.
REFERENCE: EA-AC65-9A, Page 519

5356. ANSWER #1
Almost all reciprocating engines are equipped with either a carburetor heat or an alternate-air position on the carburetor air inlet system. For starting and for ground operation, these controls should be in the cold position.
REFERENCE: EA-ITP-GB, Page 471; EA-AC65-9A, Page 490

5357. ANSWER #4
Whether in idle or takeoff thrust, the hazard area in front of a turbojet engine is 25 feet.
REFERENCE: EA-ITP-GB, Page 454; EA-AC65-9A, Page 493

5358. ANSWER #3
Carbon dioxide is the best extinguishing agent to use for a carburetor or intake fire.
REFERENCE: EA-AC65-9A, Page 491

5359. ANSWER #3
Item "D" is the proper signal for engaging a helicopter rotor.
REFERENCE: EA-ITP-GB, Page 462; EA-AC65-9A, Page 522

5360. ANSWER #4
Large radial engines have a problem of oil seeping past the piston rings and getting into the lower cylinders. This oil forms a hydraulic lock and will seriously damage the engine if some action is not taken before attempting to start the engine.
REFERENCE: EA-ITP-GB, Page 471; EA-AC65-9A, Pages 489-490

5361. ANSWER #4
Priming of most fuel injected engines is done with the mixture control in the full rich position. The engines are normally started with the mixture control in the idle cutoff position.
REFERENCE: EA-AC65-9A, Page 490

5362. ANSWER #3
Normal combustion inside an aircraft engine cylinder is started at an accurately controlled time in the operational cycle by two spark plugs. If a flake of carbon or a feather-edge on a valve is heated to incandescence, it will ignite the fuel-air mixture before the correct time, which is pre-ignition.
REFERENCE: EA-ITP-P, Page 327

5363. ANSWER #1
If the maximum allowed turbine inlet temperature is exceeded during an attempted start on a turbine engine, the start must be aborted. Overtemping of a turbine engine can cause serious damage, and in some cases requires engine overhaul.
REFERENCE: EA-ITP-GB, Page 472; EA-AC65-9A, Page 492

5364. ANSWER #1
The best way to clear a flooded engine is to crank the engine with the fuel and ignition turned off. This will purge the rich mixture from the cylinders.
REFERENCE: EA-AC65-9A, Page 491

5365. ANSWER #3
If the mixture control is not in the proper position during the starting of an engine which is internally supercharged, raw fuel can dump into the supercharger impeller and flow out the supercharger drain valve.
REFERENCE: EA-AC65-12A, Page 455

5366. ANSWER #1
Item "C" is the proper hand signal for an emergency stop.
REFERENCE: EA-ITP-GB, Page 461; EA-AC65-9A, Page 521

5367. ANSWER #1
Induction fires are best extinguished by continuing to crank the engine. If necessary, carbon dioxide can be discharged into the air intake.
REFERENCE: EA-ITP-P, Page 335; EA-AC65-9A, Page 491

5368. ANSWER #1
Immediately after an engine is started, the oil pressure should be checked.
REFERENCE: EA-ITP-GB, Page 472; EA-AC65-9A, Page 491

5369. ANSWER #3
Any time an engine is started and there is no indication of oil pressure, the engine must be shut down.
REFERENCE: EA-ITP-GB, Page 472; EA-AC65-9A, Page 492

5370. ANSWER #4
As a general rule, the mixture control should be in the full rich position when starting engines equipped with float type carburetors.
REFERENCE: EA-AC65-9A, Page 490

5371. ANSWER #2
The hazard area behind a turbojet engine when it is at idle thrust is approximately 100 feet.
REFERENCE: EA-ITP-GB, Page 454; EA-AC65-9A, Page 493

5372. ANSWER #4
A hot start is one in which ignition occurs when there is an excess of fuel and insufficient air.
REFERENCE: EA-ITP-GB, Page 472; EA-AC65-9A, Page 494

5373. ANSWER #2
When aviation gasoline is used in a turbine engine, the lead content in the fuel is deposited on the hot section components in the engine. These deposits on the turbine blades in the engine take away from the engine's efficiency.
REFERENCE: EA-ITP-P, Page 371

5374. ANSWER #4
Although aviation gasoline can be used in some turbine engines as an emergency fuel, jet fuel cannot be used in reciprocating engines. If jet fuel is inadvertently mixed with aviation gasoline, the tanks need to be drained and serviced with the proper fuel.
REFERENCE: EA-ITP-P, Page 335

5375. ANSWER #1
The viscosity of a fuel is a big factor in the fuel's susceptability to contamination. High viscosity fuel holds contaminants in suspension much more readily than does a low viscosity fuel.
REFERENCE: EA-ITP-GB, Page 469; EA-AC65-9A, Page 78

5376. ANSWER #1
100LL aviation gasoline is colored blue.
REFERENCE: EA-ITP-GB, Page 466; EA-AC65-9A, Page 76

5377. ANSWER #3
Jet fuel is a clear or straw color.
REFERENCE: EA-AC65-9A, Pages 77-78

5378. ANSWER #4
When aircraft engines grew in size and power output, fuels were demanded that had anti-detonation characteristics that were better than those of iso-octane. In order to rate these fuels, tetraethyl lead was added to increase the fuel's critical pressure. The new ratings were given performance numbers.
REFERENCE: EA-ITP-P, Page 333

5379. ANSWER #3
A scavenging agent, ethylene dibromide, is added to fuels to combine with the lead oxide and form lead bromide. This is more volatile than the oxide and it passes out the exhaust as a gas.
REFERENCE: EA-ITP-P, Page 333

5380. ANSWER #4
Although gasoline has more BTU per pound than kerosene, kerosene weighs more per gallon (volume) and therefore has more BTU per gallon.
REFERENCE: EA-ITP-P, Page 10

5381. ANSWER #3
Low vapor pressure means that the fuel does not vaporize readily and therefore would not be prone to causing a vapor lock.
REFERENCE: EA-ITP-GB, Page 467; EA-AC65-9A, Pages 77-78

5382. ANSWER #3
Detonation is a condition of uncontrolled burning which occurs inside a cylinder when the fuel/air mixture reaches its critical pressure and temperature.
REFERENCE: EA-ITP-GB, Page 381; EA-AC65-9A, Pages 74-75

5383. ANSWER #4
A fuel that is highly volatile, meaning it vaporizes readily, is very susceptible to vapor lock.
REFERENCE: EA-ITP-GB, Page 467; EA-AC65-9A, Pages 74, 77-78

5384. ANSWER #3
The numbers and letters assigned to different types of jet fuel have no relationship to the fuel's rating or performance value.
REFERENCE: EA-ITP-GB, Page 466; EA-AC65-9A, Page 77

5385. ANSWER #4
A fuel given a rating of 80/87 would indicate a rich mixture octane rating of 87 and a lean mixture octane rating of 80.
REFERENCE: EA-ITP-P, Page 333

5386. ANSWER #1
Aviation gasoline has a high heat value per pound (approximately 20,000 BTU) and is highly volatile (5.5 to 7 PSI at 100°F).
REFERENCE: EA-ITP-P, Page 332

5387. ANSWER #3
Tetraethyl lead allows engines to develop more power without detonation.
REFERENCE: EA-ITP-P, Page 333

5388. ANSWER #3
A fuel that vaporizes too readily can be a problem because of vapor lock, but a fuel that does not vaporize readily enough can cause hard starting.
REFERENCE: EA-ITP-P, Page 332; EA-AC65-9A, Page 74

5389. ANSWER #1
A dichromate treatment is a good first step prior to the painting of engine crankcases. This works well on magnesium parts.
REFERENCE: EA-AC65-12A, Page 414-415

5390. ANSWER #4
Magnesium engine parts may be cleaned using solvents, scrapers, and grit blasting. Extreme care must be taken, however, not to use a solvent or decarbonizer which will react in a harmful way with the magnesium.
REFERENCE: EA-AC65-12A, Pages 414-415

5391. ANSWER #2
Acetone is the recommended solvent for removing grease from fabric.
REFERENCE: EA-AC43.13-1A, Page 95

5392. ANSWER #2
Anodizing is a common surface treatment of aluminum alloys. When this coating is damaged in service, it can be only partially restored by chemical surface treatment.
REFERENCE: EA-AC65-9A, Page 179

5393. ANSWER #1
Boric acid or vinegar are good neutralizing agents for nickel-cadmium battery electrolyte.
REFERENCE: EA-ITP-GB, Page 135; EA-AC43.13-1A, Page 211

5394. ANSWER #1
Aliphatic naphtha is recommended for wipedown of cleaned surfaces just before painting.
REFERENCE: EA-AC65-9A, Page 189

5395. ANSWER #1
Boric acid, vinegar, or lemon juice may be used to neutralize the effects of nickel-cadmium battery electrolyte. Once the electrolyte has been neutralized, the case or drain surface should be flushed with clean water to prevent acid damage by the neutralizing agent.
REFERENCE: EA-ITP-GB, Page 135; EA-AC65-9A, Page 314

5396. ANSWER #2
Three grades of aluminum wool, coarse, medium, and fine, are used for general cleaning of aluminum surfaces.
REFERENCE: EA-AC65-9A, Page 190

5397. ANSWER #1
In addition to being used to wipe down cleaned surfaces prior to painting, aliphatic naphtha can also be used for cleaning acrylics and rubber.
REFERENCE: EA-AC65-9A, Page 189

5398. ANSWER #2
Methyl ethyl ketone is available as a solvent cleaner for metal surfaces and paint stripper for small areas.
REFERENCE: EA-AC65-9A, Page 189

5399. ANSWER #3
Chemical cleaners must be used with great care in cleaning assembled aircraft. The danger of entrapping corrosive materials in fraying surfaces and crevices counteracts any advantages in their speed and effectiveness.
REFERENCE: EA-AC65-9A, Page 190

5400. ANSWER #4
Caustic cleaning products, such as alkaline cleaners, are corrosive when in contact with aluminum.
REFERENCE: EA-AC65-9A, Pages 185-190

5401. ANSWER #4
Light oiling and retightening will help keep fretting corrosion to a minimum.
REFERENCE: EA-AC43-4, Item 49

5402. ANSWER #2
Very severe intergranular corrosion may sometimes cause the surface of a metal to "exfoliate".
REFERENCE: EA-ITP-GB, Page 405; EA-AC65-9A, Page 172

5403. ANSWER #1
On copper and copper alloys, corrosion forms a greenish film.
REFERENCE: EA-AC65-9A, Page 171

5404. ANSWER #4
Alodining is a simple chemical treatment for all aluminum alloys to increase their corrosion resistance and to improve their paint-bonding qualities.
REFERENCE: EA-ITP-GB, Page 419; EA-AC65-9A, Page 183

5405. ANSWER #4
All metals and alloys are electrically active and have a specific electrical potential in a given chemical environment. Two different metals in contact with each other can create a corrosive environment.
REFERENCE: EA-ITP-GB, Pages 405-406; EA-AC65-9A, Page 172

5406. ANSWER #1
The most practicable means of controlling the corrosion of steel is the complete removal of corrosion products by mechanical means and restoring the corrosion-preventive coatings.
REFERENCE: EA-ITP-GB, Pages 420-421; EA-AC65-9A, Page 177

5407. ANSWER #2
After a thorough rinsing, alodine is applied by dipping, spraying or brushing.
REFERENCE: EA-ITP-GB, Page 419; EA-AC65-9A, Page 183

5408. ANSWER #4
Exfoliation is a severe case of intergranular corrosion.
REFERENCE: EA-ITP-GB, Page 405; EA-AC65-9A, Page 172

5409. ANSWER #1
Alodining leaves a film of hydroxide on the surface of aluminum when it is wet, and dries to a hard oxide finish to help prevent corrosion.
REFERENCE: EA-AC65-9A, Page 183

5410. ANSWER #1
The sodium and potassium nitrate bath leaves a corrosive substance on parts which must be thoroughly rinsed off in a hot water application.
REFERENCE: EA-AC65-9A, Page 214

5411. ANSWER #4
Intergranular corrosion is very difficult to detect in its original stages. Ultrasonic and eddy current inspection methods are being used with a great deal of success.
REFERENCE: EA-ITP-GB, Page 405; EA-AC65-9A, Page 172

5412. ANSWER #1
The lack of uniformity in the structure of metals which makes them particularly susceptible to intergranular corrosion occurs in the metal during the heating and cooling process.
REFERENCE: EA-ITP-GB, Page 405; EA-AC65-9A, Page 172

5413. ANSWER #1
Electrolytic corrosion between two dissimilar metals is often under the surface of the skin and may not be visible.
REFERENCE: EA-AC65-9A, Page 172

5414. ANSWER #3
Corroded magnesium may generally be treated as follows: (1) clean and strip the paint from the area to be treated, and (2) using a stiff, hog-bristle brush, break loose and remove as much of the corrosion products as practicable.
REFERENCE: EA-AC65-9A, Page 179

5415. ANSWER #4
After the final spraying of corrosion preventive, the crankshaft must not be moved or the seal of the mixture will be broken.
REFERENCE: EA-AC65-12A, Page 387

5416. ANSWER #2
Before applying soap and water to plastic surfaces, they should be flushed with fresh water to dissolve salt deposits and wash away dust particles.
REFERENCE: EA-AC65-9A, Page 185

5417. ANSWER #2
Surface oil, hydraulic fluid, grease or fuel can be removed from aircraft tires by washing with a mild soap solution.
REFERENCE: EA-AC65-9A, Page 186

5418. ANSWER #2
The best way to keep dissimilar metal corrosion to a minimum is to keep the metals separated by a coating of paint or zinc chromate primer.
REFERENCE: EA-ITP-GB, Page 419; EA-AC65-9A, Page 172

5419. ANSWER #2
Extensive pitting damage may result from contact between dissimilar metal parts in the presence of a conductor. This corrosion is referred to as an electrochemical attack.
REFERENCE: EA-ITP-GB, Pages 405-406

5420. ANSWER #4
To prevent corrosion between dissimilar metal joints in which magnesium alloy is involved, each surface is insulated as follows: At least two coats of zinc chromate are applied to each surface. Next, a layer of pressure-sensitive vinyl tape 0.003 inch thick is applied smoothly and firmly enough to prevent air bubbles and wrinkles.
REFERENCE: EA-AC65-9A, Page 180

5421. ANSWER #2
To protect the interior of structural steel and aluminum tubing against corrosion, coat the tube interior by flushing with hot linseed oil.
REFERENCE: EA-AC43.13-1A, Page 126

5422. ANSWER #3
Aluminum alloy must be immediately quenched in water to solidify the alloy elements into extremely tiny grains. If this is not done, the corrosion resistance will be impaired.
REFERENCE: EA-ITP-GB, Page 405; EA-AC65-9A, Pages 213-214

5423. ANSWER #4
1,000,000 equals 10 to the sixth power, or six values of ten multiplied together.
REFERENCE: EA-ITP-GB, Pages 10-11; EA-AC65-9A, Page 14

5424. ANSWER #1
Finding the square root of 1,746 means finding the number, which multiplied by itself, equals 1,746. This number is 41.7852.
REFERENCE: EA-ITP-GB, Page 27; EA-AC65-9A, Page 13

5425. ANSWER #4
Nine raised to the fourth power means four values of nine multiplied together. This number is 6,561.
REFERENCE: EA-ITP-GB, Page 27; EA-AC65-9A, Page 13

5426. ANSWER #4
The square root of 3722.1835 means what number, when multiplied by itself, will equal 3722.1835. This number is 61.0097.
REFERENCE: EA-ITP-GB, Page 27; EA-AC65-9A, Page 13

5427. ANSWER #3
Finding the square of 212 means multiplying 212 by itself. The square of 212 is 44,944.
REFERENCE: EA-ITP-GB, Page 27; EA-AC65-9A, Page 13

5428. ANSWER #1
Ten to the negative sixth power means one divided by ten to the positive sixth power. The answer is 0.000001.
REFERENCE: EA-ITP-GB, Pages 10-11; EA-AC65-9A, Page 14

5429. ANSWER #1
First take the square root of four, or 2. Two raised to the fifth power means five values of 2 multiplied together, or 32.
REFERENCE: EA-ITP-GB, Page 27; EA-AC65-9A, Page 13

5430. ANSWER #4
3.47 times 10 to the negative fourth power means 3.47 times 1 divided by 10 to the positive fourth power. ($3.47 \times 1 \div 10^4$).
REFERENCE: EA-ITP-GB, Pages 10-11; EA-AC65-9A, Pages 14-15

5431. ANSWER #1
1.63 times 10 to the fourth power equals 16,300 because 10 to the fourth power is four values of ten multiplied together, or 10,000. 1.63 times 10,000 is 16,300.
REFERENCE: EA-ITP-GB, Pages 10-11; EA-AC65-9A, Pages 13-14

5432. ANSWER #3
The square root of 125.131 is the number, which multiplied by itself, is equal to 125.131.
REFERENCE: EA-ITP-GB, Pages 10-11, 27; EA-AC65-9A, Page 13

5433. ANSWER #4
The square root of 16 is 4. Four to the fourth power (four 4s multiplied together) is 256.
REFERENCE: EA-ITP-GB, Page 27; EA-AC65-9A, Page 13

5434. ANSWER #4
The square root of 31 is 5.567 and the square root of 43 is 6.557. The sum of these two is 12.124, divided by 17 squared (289), which equals .0419.
REFERENCE: EA-ITP-GB, Page 27; EA-AC65-9A, Page 13

5435. ANSWER #1
Seven raised to the third power is three 7s multiplied together, or 343. The square root of 39 (the number multiplied by itself which would equal 39) is 6.24. These two numbers added together total 349.24.
REFERENCE: EA-ITP-GB, Page 27; EA-AC65-9A, Page 13

5436. ANSWER #4
The square root of 1,824 (the number multiplied by itself which would equal 1,824) is 42.708. This number is equal to .42708 times ten to the second power.
REFERENCE: EA-ITP-GB, Page 27; EA-AC65-9A, Pages 13-14

5437. ANSWER #3
Piston displacement is a volume measurement and it deals with the total volume of all the engine's cylinders.
REFERENCE: EA-ITP-GB, Page 34; EA-AC65-9A, Page 23

5438. ANSWER #2
The area of a trapezoid is equal to 1/2 the sum of the two parallel sides (bases) multiplied by the height. One half of the two bases is 8, multiplied by the height of 4 gives 32 square feet.
REFERENCE: EA-ITP-GB, Page 31; EA-AC65-9A, Page 19

5439. ANSWER #3
One side of the sheet metal needs to be 20 inches in length to accomodate the cylinder's length. The distance around the cylinder, or circumference, will be the other side. Circumference is equal to 3.1416 multiplied by diameter, or 25^9/$_{64}$ inches.
REFERENCE: EA-ITP-GB, Page 32; EA-AC65-9A, Page 21

5440. ANSWER #2
The area of a triangle is equal to 1/2 the altitude multiplied by the base. In this case, 1/2 of 3 times 4, which equals 6 square inches.
REFERENCE: EA-ITP-GB, Page 30; EA-AC65-9A, Page 18

5441. ANSWER #1
Pascal's law states that force is equal to pressure times area (F = PA). In this case there is a pressure of 850 PSI and an area of 1.2 square inches, so the force would be 1,020 pounds.
REFERENCE: EA-ITP-GB, Pages 80-81; EA-AC65-9A, Pages 232-233

5442. ANSWER #4
The piston displacement of one cylinder is the volume displaced by the piston in that cylinder. Volume is equal to the radius of the cylinder squared multiplied by 3.1416 and also multiplied by the height of the cylinder. The radius of this cylinder is one half of the bore (diameter) and the height is equal to the stroke of the piston.
REFERENCE: EA-ITP-GB, Pages 34-35; EA-AC65-9A, Page 23

5443. ANSWER #1
The volume of a rectangular shaped box is equal to its length, multiplied by its width, multiplied by its height. This box is 60 by 30 by 12, or 21,600 cubic inches. To convert this number to cubic feet, it must be divided by the number of cubic inches in a cubic foot, or 1728. This gives us an answer of 12.5 cubic feet.
REFERENCE: EA-ITP-GB, Page 34; EA-AC65-9A, Page 21

5444. ANSWER #3
Given that 7.5 GAL = 1 CU FT, 60 gallons divided by 7.5 equals 8.0 cubic feet. No further reference required.

5445. ANSWER #4
The area of a trapezoid is equal to one half of the sum of the two parallel sides (bases) multiplied by the height. One half of the sum of the bases is 5, multiplied by the height of 2 gives an area of 10 square feet.
REFERENCE: EA-ITP-GB, Page 31; EA-AC65-9A, Page 19

5446. ANSWER #3
The area of the triangle being asked for in this question is one-half of the area of the rectangle, for which we know the length (16.8) and width (7.5). The area of the rectangle (7.5 × 16.8) divided by 2 is the area of the triangle, or 63 square inches.
REFERENCE: EA-ITP-GB, Page 30; EA-AC65-9A, Pages 16-18

5447. ANSWER #3
The displacement of a piston (volume) is equal to the area of the piston (1/2 its diameter squared multiplied by 3.1416) multiplied by the stroke of the piston. For this piston, the displacement is 7.0685 cubic inches.
REFERENCE: EA-ITP-GB, Page 34; EA-AC65-9A, Page 23

5448. ANSWER #2
The first thing to do here is find the volume of the tank. The volume of a rectangular shaped tank is equal to its length, multiplied by its width, multiplied by its depth. Its volume would be 10 cubic feet, and since 7.5 gallons fit in one cubic foot, the tank would hold 75 gallons.
REFERENCE: EA-ITP-GB, Page 34; EA-AC65-9A, Page 21

5449. ANSWER #3
The volume of a rectangular shaped tank is equal to its length, multiplied by its width, multiplied by its depth. The volume of this tank would be 2041.875 cubic inches. Since one gallon fits in 231 cubic inches, the tank would hold 8.8 gallons.
REFERENCE: EA-ITP-GB, Page 34; EA-AC65-9A, Page 21

5450. ANSWER #1

The volume of a cylinder (displacement) is equal to the radius of the cylinder squared multiplied by 3.1416, and again multiplied by the height of the cylinder. The radius in this question would be 1/2 of the cylinder bore and the height of the cylinder. The radius in this question would be 1/2 of the cylinder bore and the height would be the stroke of the piston (the difference between 8.5 and 4). The approximate displacement of the engine in question would be 200 cubic inches.
REFERENCE: EA-ITP-GB, Page 34; EA-AC65-9A, Page 23

5451. ANSWER #3

The volume of a rectangular tank is equal to its length, multiplied by its width, multiplied by its depth. In this case the volume would be 4331.25 cubic inches.
REFERENCE: EA-ITP-GB, Page 34; EA-AC65-9A, Page 21

5452. ANSWER #4

The decimal fraction .020 would be stated verbally as two one-hundreths. 2/100 would equal 1/50.
REFERENCE: EA-ITP-GB, 1, Page 21; EA-AC65-9A, Page 7

5453. ANSWER #1

The fraction 7/32 is converted to a decimal by dividing 32 into 7. Seven divided by 32 equals .21875, so 1 and 7/32 would equal 1.21875.
REFERENCE: EA-ITP-GB, Page 20; EA-AC65-9A, Page 8

5454. ANSWER #3

The compression ratio of a cylinder is the ratio of cylinder volume when the piston is at the bottom of its stroke and cylinder volume when the piston is at the top of its stroke. In this case the compression ratio is 84 to 14 (84-70) or 6 to 1.
REFERENCE: EA-ITP-GB, Page 25; EA-AC65-9A, Page 10

5455. ANSWER #4

The common fraction 7/8 would equal .875 as a decimal fraction (7 ÷ 8). To convert a decimal fraction to a percent the decimal point is moved two places to the right, or 87.5%.
REFERENCE: EA-ITP-GB, Page 24; EA-AC65-9A, Page 9

5456. ANSWER #4

There is an inverse or indirect relationship between the speed of a gear and the number of teeth on the gear. This means that when two gears are meshed together, the gear with the greater number of teeth will turn at the slower speed. The speed it will turn is the fractional value obtained by dividing the number of teeth on the smaller gear by the number of teeth on the larger gear. In this case, the speed would be 14/42 times 420, or 140 RPM.
REFERENCE: EA-ITP-GB, Page 26; EA-AC65-9A, Page 11

5457. ANSWER #1

If 87% of the engine's power equals 108 horsepower, then 100% of the engine's power would be 124 horsepower. Sixty-five percent of 124 horsepower would be 80 horsepower.
REFERENCE: EA-ITP-GB, Page 24; EA-AC65-9A, Page 9

5458. ANSWER #1

To convert the common fraction 43/32 to a decimal fraction, 43 is divided by 32. The answer obtained is 1.34375.
REFERENCE: EA-ITP-GB, Page 20; EA-AC65-9A, Page 8

5459. ANSWER #4

The decimal fraction 0.09375, or 9375 ÷ 100000, would reduce down to 3/32.
REFERENCE: EA-ITP-GB, Page 21; EA-AC65-9A, Page 7

5460. ANSWER #4

The common fraction 5/8 would equal .625 (5 ÷ 8). The decimal .625 is converted to a percent by moving the decimal two places to the right, or 62.5%.
REFERENCE: EA-ITP-GB, Page 24; EA-AC65-9A, Page 9

5461. ANSWER #3

To convert the common fraction 77/64 to a decimal fraction, 77 is divided by 64. The answer to this division is 1.2031.
REFERENCE: EA-ITP-GB, Page 20; EA-AC65-9A, Page 8

5462. ANSWER #3

The first thing to find out is how many miles the airplane is flying per gallon. This is solved by dividing the number of miles flown (875) by the number of gallons used (70). It works out to 12.5 miles per gallon. 3,000 miles divided by 12.5 works out to 240 gallons needed to fly the distance.
REFERENCE: EA-ITP-GB, Page 26; EA-AC65-9A, Page 11

5463. ANSWER #3

The speed ratio between gears is the inverse of the teeth ratio between gears. In this case, the teeth ratio is 36 to 20, so the speed ratio would be 20 to 36. This ratio would be 5 to 9, which is not a possible given choice. The answer 9 to 5 would seem to be the intended choice, with the digits in the wrong order.
REFERENCE: EA-ITP-GB, Page 25; EA-AC65-9A, Page 10

5464. ANSWER #3
The relationship between the speed of the two gears which are meshed together is the inverse of the relationship between the number of teeth on the gears. This means the gear with the greater number of teeth will turn at the slowest speed. The ratio between the number of teeth in this problem is 14 to 42; the inverse of which is 42 to 14. The small gear (pinion) would have a $^{42}/_{14}$ greater speed, or 420 RPM.
REFERENCE: EA-ITP-GB, Page 26; EA-AC65-9A, Page 11

5465. ANSWER #4
If the parts department's profit is 12%, then the cost of the part (100%) plus the 12% profit, would need to equal the selling price of $145.60. One-hundred twelve percent of 130 dollars is equal to 145.60.
REFERENCE: EA-ITP-GB, Page 24; EA-AC65-9A, Page 9

5466. ANSWER #3
65% of 125 horsepower is 81.25 horsepower (.65 × 125).
REFERENCE: EA-ITP-GB, Page 24; EA-AC65-9A, Page 9

5467. ANSWER #3
75% of 98 horsepower is 73.5 horsepower (.75 × 98).
REFERENCE: EA-ITP-GB, Page 24; EA-AC65-9A, Page 9

5468. ANSWER #1
The decimal fraction 0.17187 is 17187 ÷ 100000. This common fraction is most nearly equal to $^{11}/_{64}$.
REFERENCE: EA-ITP-GB, Page 21; EA-AC65-9A, Page 7

5469. ANSWER #3
To convert the common fraction $^{31}/_{64}$ to a decimal fraction, 31 is divided by 64. This gives a decimal value of .48437.
REFERENCE: EA-ITP-GB, Page 20; EA-AC65-9A, Page 8

5470. ANSWER #2
To find out what percent 65 engines are of 80 engines, 65 is divided by 80. This gives a value of .8125, which can be converted to a percent by moving the decimal two places to the right. The answer would be 81.25%.
REFERENCE: EA-ITP-GB, Page 24; EA-AC65-9A, Page 9

5471. ANSWER #3
The common fraction $^{7}/_{32}$ is converted to a decimal fraction by dividing 7 by 32. This would give an answer of .4375.
REFERENCE: EA-ITP-GB, Page 20; EA-AC65-9A, Page 8

5472. ANSWER #1
This question is asking what percent 635.30 is of 900 hours. To arrive at this answer, 635.30 is divided by 900, which gives .7058. This is converted to a percent by moving the decimal two places to the right, which gives 70.58%.
REFERENCE: EA-ITP-GB, Page 24; EA-AC65-9A, Page 9

5473. ANSWER #1
To solve this problem, it is necessary to convert the 10 feet to inches, or 120 inches. The ratio then becomes 120 to 30, or 4 to 1.
REFERENCE: EA-ITP-GB, Page 25; EA-AC65-9A, Page 10

5474. ANSWER #1
To answer this question, it is first necessary to know that one horsepower is equal to 746 WAtts. One-half horsepower would equal 373 Watts, but this motor is only 85% efficient, so it would draw 438.8 Watts (373 ÷ .85). The current draw is equal to the Watts divided by the voltage, or 14.6 amps.
REFERENCE: EA-ITP-GB, Pages 107-108, 24; EA-AC65-9A, Pages 282-284

5475. ANSWER #3
To solve this problem, it is very important that the proper order of operations be followed. The values in parentheses must be solved first, then added to each other, and finally divided by 2.
Step 1: $4(-3) = -12$
Step 2: $-9(2) = -18$
Step 3: $(-12) + (-18) = -30$
Step 4: $-30 \div 2 = -15$
REFERENCE: EA-ITP-GB, Page 23; EA-AC65-9A, Page 12

5476. ANSWER #3
The operation 64 times $^{3}/_{8}$ equals 24, and 24 divided by $^{3}/_{4}$ equals 32.
REFERENCE: EA-ITP-GB, Pages 17, 19; EA-AC65-9A, Page 4

5477. ANSWER #4
This problem can be solved by the following steps:
Step 1: $-18.4 + .07 = -18.33$
Step 2: $-18.33 - 2.2 = -20.53$
Step 3: $-20.53 + 8.36 = -12.17$
REFERENCE: EA-ITP-GB, Page 23; EA-AC65-9A, Page 12

5478. ANSWER #2
To solve this problem it is first necessary to find the least common denominator (LCD) for the fractions. The LCD is 24ths. Converted to the LCD, the problem would read: $12/24 + 21/24 - 8/24 = 25/24$ or $11/24$.
REFERENCE: EA-ITP-GB, Pages 12-13; EA-AC65-9A, Page 3

5479. ANSWER #3
This problem can be solved by the following steps:
Step 1: $11/2 \times 7/8$ can be changed to $3/2 \times 7/8$, which equals $21/16$
Step 2: $1/4$ is equal to $4/16$
Step 3: $4/16 + 21/16 = 19/16$
REFERENCE: EA-ITP-GB, Pages 12-13, 17; EA-AC65-9A, Pages 3-4

5480. ANSWER #4
32 times $3/8$ equals $96/8$ divided by $1/16$ equals 72.
REFERENCE: EA-ITP-GB, Pages 17, 19; EA-AC65-9A, Page 4

5481. ANSWER #2
To answer this question, it is first necessary to convert the gallons to pounds. 200 gallons would weigh 1200 pounds, so the ratio is 1200 to 1680, or 5 to 7.
REFERENCE: EA-ITP-GB, Page 25; EA-AC65-9A, Page 10

5482. ANSWER #1
$21/2$ times $13/8$ should be converted to $5/2$ times $11/8$. The product of these two numbers is $55/16$. $55/16$ plus $23/4$ equals $63/16$.
REFERENCE: EA-ITP-GB, Pages 13, 17; EA-AC65-9A, Pages 3-4

5483. ANSWER #1
It is first necessary to add 41 and 40, which equals 81. Eighty-one times $5/9$ equals 45. Forty-five minus 40 equals 5.
REFERENCE: EA-ITP-GB, Page 17; EA-AC65-9A, Page 4

5484. ANSWER #2
32 times $3/16$ equals 6. Six divided by $1/2 = 12$.
REFERENCE: EA-ITP-GB, Pages 17, 19; EA-AC65-9A, Page 4

5485. ANSWER #2
To solve this problem, it is first necessary to add 30 and 34, which would equal 64. Sixty-four times $2/4$ would equal 32, and 32 times 5 would equal 160.
REFERENCE: EA-ITP-GB, Page 17; EA-AC65-9A, Page 4

5486. ANSWER #4
To solve this problem, solve the top line (numerator position) first. This is done as follows:
Step 1: $-35 + 25 = -10$
Step 2: $-10(-7) = +70$
Step 3: $3.1416(16^{-2}) = 3.1416(1/16^2) = 3.1416 \quad (1/256) = .01227$
Step 4: $70 + .01772 = 70.01227$ (This is a solution to the top line)
Step 5: 70.01227 divided by the square root of 25, or 5, equals 14.002.
REFERENCE: EA-ITP-GB, Pages 23-28; EA-AC65-9A, Chapter 1

5487. ANSWER #2
This problem can be solved by the following steps:
Step 1: $125 \div -4 = -31.25$
Step 2: $-36 \div -6 = 6$
Step 3: $-31.25 \div 6 = -5.20$
REFERENCE: EA-ITP-GB, Page 23; EA-AC65-9A, Page 12

5488. ANSWER #1
To solve this problem, the values in parentheses must be solved first. Then the value in the bracket is solved. The last step is to multiply the bracket value by the value outside the bracket, and then add. The steps are as follows:
Step 1: $2 + 3 = 5$
Step 2: $5 \times -6 = -30$
Step 3: $-30 + 4 = -26$
Step 4: $-26 \times -3 = +78$
Step 5: $+78 + 4 = 82$
REFERENCE: EA-ITP-GB, Page 23; EA-AC65-9A, Page 12

5489. ANSWER #4
To solve this problem, the values in parentheses must be solved first. Then the value in the bracket is solved. The last step is to multiply the bracket value by the value outside the bracket, and then subtract. The steps are as follows:
Step 1: $-8 + 4 = -4$
Step 2: $-9 \times -4 = +36$
Step 3: $7 + 3 = 10$

Step 4: $-2 \times 10 = -20$
Step 5: $36 - 20 = 16$
Step 6: $16 \times -6 = -96$
REFERENCE: EA-ITP-GB, Page 23; EA-AC65-9A, Page 12

5490. ANSWER #1
To solve this problem, the values in parentheses must be solved first. Then the value in the bracket is solved. The last step is to multiply the bracket value by the value outside the bracket, and then subtract. The steps are as follows:
Step 1: $3 - 4 = -1$
Step 2: $4 - (-1)$, or $4 + 1 = 5$, and $5 + 13 = 18$
Step 3: $18 \times 5 = 90$
Step 4: $90 - 6 = 84$
REFERENCE: EA-ITP-GB, Page 23, EA-AC65-9A, Page 12

5491. ANSWER #3
The aircraft maintenance records provide a place for indicating compliance with Airworthiness Directives.
REFERENCE: EA-ITP-GB, Page 541; EA-AC65-9A, Page 463

5492. ANSWER #1
In order to fly an aircraft that has not been returned to service after an annual inspection, the operator must receive a special flight permit.
REFERENCE: FAR Part 91.169

5493. ANSWER #3
If an overweight or hard landing has, or is suspected to have occurred, the mechanic must give the aircraft a careful inspection to ascertain whether or not any damage has been done.
REFERENCE: EA-ITP-GB, Pages 528 and 529; EA-AC65-9A, Page 463

5494. ANSWER #3
When a mechanic finds discrepancies that render an aircraft unairworthy, he must provide the owner with a list of those discrepancies.
REFERENCE: FAR Part 43.11

5495. ANSWER #1
Airworthiness Directives are the the media used by the FAA to notify owners and other interested persons of unsafe conditions, and under what conditions that the product may continue to be operated.
REFERENCE: EA-AC65-9A, Page 465

5496. ANSWER #3
When major repair is made to an airframe, a Form 377 must be filled out to record the work that has been done.
REFERENCE: EA-ITP-GB, Page 542

5497. ANSWER #2
Crankcase separation of a reciprocating engine equipped with an integral supercharger is considered a powerplant major repair.
REFERENCE: FAR Part 43, Appendix A

5498. ANSWER #4
Repair of portions of the skin sheets, by making additional seams, is considered a major airframe repair.
REFERENCE: FAR Part 43, Appendix A

5499. ANSWER #4
A sudden stoppage inspection will be done whenever an aircraft powerplant has been subjected to sudden deceleration or stoppage.
REFERENCE: EA-AC43.13-1A, Paragraph 679

5500. ANSWER #4
According to AC43.13-1A, this crack could be repaired by cutting out the defective area, installing a flush repair using the same material, rivet size, and unit spacing as the adjacent area and a rivet edge distance of two times the diameter of the rivets used.
REFERENCE: EA-AC43.13-1A, Page 59

5501. ANSWER #1
To repair a plastic side window, a piece of plastic should be cut of sufficient size to cover the damaged area and extend at least 3/4 inch on each side of the crack or hole. If the patch is to be shaped, heat it in an oil bath at a temperature of 248-302° Fahrenheit. The path is then shaped, adhesive is applied and the patch is put in place with a pressure of 5-10 PSI applied for a minimum of 3 hours.
REFERENCE: EA-AC43.13-1A, Pages 160-161

5502. ANSWER #3
Repair dents at a steel-tube cluster-joint by welding a specially formed steel patch plate over the dented area and surrounding tubes.
REFERENCE: EA-AC43.13-1A, Page 32

5503. ANSWER #3
The person who approves or disapproves for return to service after an inspection is the one responsible for making the record entry.
REFERENCE: EA-FAR 43.11

5504. ANSWER #2
If a certificated mechanic is appropriately rated, then #2 would be the correct answer.
REFERENCE: No specific reference

5505. ANSWER #3
Any time a repair is entered in the aircraft logbook, the name of the person who performed the work and the date the job was completed must be entered in the record.
REFERENCE: EA-FAR 43.9

5506. ANSWER #3
Minor repairs require only an entry in the aircraft's records. If it were a major repair, it would also require a Form 337.
REFERENCE: EA-FAR 43.9

5507. ANSWER #2
According to FAR 43, Appendix B, when a major repair is performed a Form 337 must be made out in duplicate. One copy of the 337 goes with the aircraft records and the other copy goes to the FAA.
REFERENCE: EA-FAR 43, Appendix B

5508. ANSWER #4
The person ultimately responsible for maintaining an aircraft in an airworthy condition, including its maintenance records, is the aircraft owner.
REFERENCE: EA-FAR 91.173

5509. ANSWER #1
The minimum items to be included in a 100-hour or annual inspection are contained in Appendix D of FAR 43.
REFERENCE: EA-FAR 43, Appendix D

5510. ANSWER #2
FAA Form 337 is used to record and document major repairs and major alterations.
REFERENCE: EA-FAR 43, Appendix B

5511. ANSWER #2
Any time an aircraft is inspected, the proper entries must be made in the aircraft records before the aircraft can be returned to service.
REFERENCE: EA-FAR 43.11

5512. ANSWER #3
On the back of a Form 337, a description of the work accomplished is entered in the same manner as work accomplished is entered in the aircraft maintenance record.
REFERENCE: EA-ITP-GB, Pages 541-545

5513. ANSWER #1
Appendix D of FAR 43 contains the minimum scope and detail of items to be inspected on a 100-hour or annual inspection.
REFERENCE: EA-FAR 43, Appendix D

5514. ANSWER #1
Certified Airframe and Powerplant mechanics are authorized to perform and return to service an aircraft after a 100-hour inspection, provided they are current and have the necessary experience, equipment, and technical data.
REFERENCE: EA-FAR 65.85 and 65.87

5515. ANSWER #2
According to Pascal's Law, the force available in a hydraulic system is equal to the pressure in the system multiplied by the cross sectional area of the piston.
REFERENCE: EA-ITP-GB, Pages 80-81; EA-AC65-9A, Pages 232-233

5516. ANSWER #
Pressure is an important factor that effects changes in the state of matter. The higher the pressure, the higher the boiling point of water (the temperature when water changes from a liquid to a gas).
REFERENCE: EA-AC65-9A, Page 221

5517. ANSWER #4
33,000 foot pounds of work per minute, or 550 foot pounds of work per second, equals one horsepower.
REFERENCE: EA-ITP-GB, Page 56; EA-AC65-9A, Page 252

5518. ANSWER #3
Convection, conduction, and radiation are all methods of heat transfer.
REFERENCE: EA-ITP-GB, Pages 70-71; EA-AC65-9A, Pages 257-259

5519. ANSWER #1
Although the information is given, the weight of the engine and the height to which it is raised has nothing to do with this question. All that matters is the distance the assembly is moved, and how much force is required to make the move. Work equals force multiplied by distance, or 70 times 12. Eight hundred forty foot pounds of work input is needed for the move.
REFERENCE: EA-ITP-GB, Page 55; EA-AC65-9A, Page 249

5520. ANSWER #3
The actual amount of water vapor in a mixture of air and water, expressed in grams per cubic meter or pounds per cubic foot, is absolute humidity.
REFERENCE: EA-AC65-9A, Page 239

5521. ANSWER #3
The greater the pressure differential between metered and unmetered pressure, the greater the rate of flow through a metering orifice.
REFERENCE: EA-ITP-GB, Pages 82-83; EA-AC65-9A, Page 236

5522. ANSWER #4
Vapor pressure is that portion of atmospheric pressure exerted by the moisture in the air.
REFERENCE: EA-AC65-9A, Page 240

5523. ANSWER #2
The block and tackle in the figure has four supporting ropes, which gives it a mechanical advantage of 4. With a mechanical advantage of four, 15 pounds of effort would lift 60 pounds.
REFERENCE: EA-ITP-GB, Page 60; EA-AC65-9A, Page 249

5524. ANSWER #4
The greater the amount of water vapor in the air, the less the air will weigh. A sample of air which is 50% water vapor will weigh less than the other choices in this question.
REFERENCE: EA-AC65-9A, Page 239

5525. ANSWER #2
Relative humidity is the ratio of the water vapor actually present in the atmosphere to the amount that would be present if the air were saturated.
REFERENCE: EA-AC65-9A, Page 239

5526. ANSWER #4
Power is equal to force times distance, divided by time. One hundred twenty pounds of force, exerted over 20 feet in 20 seconds would expend 120 foot pounds per second of power.
REFERENCE: EA-ITP-GB, Page 56; EA-AC65-9A, Page 252

5527. ANSWER #4
Boyle's Law states that the volume of an enclosed dry gas varies inversely with the absolute pressure, provided the temperature remains constant. If the volume of a gas is doubled, the pressure will be cut in half.
REFERENCE: EA-ITP-GB, Page 75; EA-AC65-9A, Page 228

5528. ANSWER #4
Boyle's Law applies to conditions exerted on confined gases. Because the substance in question is a confined liquid, there is no change in its density.
REFERENCE: EA-ITP-GB, Page 75; EA-AC65-9A, Page 228

5529. ANSWER #1
The density of a gas increases in direct proportion to the pressure exerted on it.
REFERENCE: EA-AC65-9A, Page 223

5530. ANSWER #3
Whether raising or lowering, the work input needed to move a weight a specified distance is equal to the force multiplied by the distance. Vertical moves of a weight require a force equal to the object's weight. In this question a 120 pound weight is being moved a distance of 3 feet, so a work input of 360 pounds is needed.
REFERENCE: EA-ITP-GB, Page 55; EA-AC65-9A, Pages 249-250

5531. ANSWER #4
The true landing speed of an aircraft is going to be greatest when the air density is the lowest, because with less air density there is less lift on the wing. High temperature and high humidity cause a decrease in air density.
REFERENCE: EA-AC65-9A, Pages 223, 239

5532. ANSWER #3
The size of the line has no effect on the answer to this question. Pascal's Law states that the force available at a piston is equal to the pressure in the system multiplied by the cross sectional area of the piston (F = PA). 800 PSI of system pressure multiplied by a 10 square inch piston would give an output force of 8,000 pounds.
REFERENCE: EA-ITP-GB, Pages 80-81; EA-AC65-9A, Pages 232-233

5533. ANSWER #4
Power is equal to force, multiplied by distance, and divided by time.
REFERENCE: EA-ITP-GB, Page 56; EA-AC65-9A, Page 252

5534. ANSWER #1
By using an inclined plane, it is possible to raise an object a vertical distance without supplying an input force equal to the object's weight. If an inclined plane is used which is 3 times longer than the vertical height the object is being raised, then only 1/3 as much force is needed (disregarding friction). In this question, the inclined plane is 9 feet long and the vertical height is 3 feet. Under these conditions, 120 pounds could be raised by using an input force of 40 pounds.
REFERENCE: EA-ITP-GB, Page 59; EA-AC65-9A, Page 247

5535. ANSWER #3
Heat is a form of energy that causes molecular agitation within a material. The amount of this agitation can be measured by a scale known as temperature.
REFERENCE: EA-ITP-GB, Page 71; EA-AC65-9A, Page 225

5536. ANSWER #3
Absolute humidity is the actual amount of water vapor present in a mixture of air and water. It is usually measured in grams per cubic meter or pounds per cubic foot.
REFERENCE: EA-AC65-9A, Page 239

5537. ANSWER #2
The size of the feed line has no effect on the answer to this question. According to Pascal's Law, the force available at a piston is equal to the pressure in the system multiplied by the cross sectional area of the piston. 2,000 PSI in this system is acting on a 10 square inch piston, which would produce an output force of 20,000 pounds.
REFERENCE: EA-ITP-GB, Pages 80-81; EA-AC65-9A, Pages 232-233

5538. ANSWER #1
Dew point is the temperature to which humid air must be cooled at constant pressure to become saturated. As air is cooled, it is not able to hold as much water vapor in suspension. When it is cooled to the point that it can no longer hold the water vapor, the water vapor condenses and is deposited on the ground as dew.
REFERENCE: EA-AC65-9A, Page 240

5539. ANSWER #1
According to the General Gas Law (a combination of Boyle's and Charles' Laws), the volume of a gas, multiplied by its absolute pressure and divided by its absolute temperature, will, if one of the conditions is allowed to change, equal the new volume of the gas, multiplied by its new absolute pressure and divided by its new absolute temperature. If both the volume and the absolute temperature are doubled, their effects will cancel out and the pressure of the gas will not change.
REFERENCE: EA-ITP-GB, Page 76; EA-AC65-9A, Pages 229-230

5540. ANSWER #4
The FAA issues airworthiness directives whenever they feel an unsafe condition exists in a product, and the same condition is likely to exist or develop in other products of the same type design.
REFERENCE: EA-FAR 39.1

5541. ANSWER #4
Supplemental type certificates (STCs) are not patents. More than one person can apply for the same design change and be issued an STC, if the applicable airworthiness requirements are met. Just because an item is manufactured in accordance with the Technical Standard Order system does not mean that it can be installed in an aircraft without further approval.
REFERENCE: EA-FAR 21.113 and 21.115; EA-FAR 21, Subpart O

5542. ANSWER #3
Since January of 1958, aircraft specifications have been contained in documents called Type Certificate Data Sheets.
REFERENCE: EA-ITP-GB, Page 474; EA-AC65-9A, Page 465

5543. ANSWER #4
Airworthiness Directives are issued by the Federal Aviation Administration.
REFERENCE: EA-AC65-9A, Page 465

5544. ANSWER #4
Of the items mentioned in this question, the only one you would find in a Type Certificate Data Sheet is the location of the datum.
REFERENCE: EA-ITP-GB, Page 478; EA-AC65-9A, Page 466

5545. ANSWER #3
The Type Certificate Data Sheet for an airplane lists the approved engines and propellers that can be installed on the airplane.
REFERENCE: EA-ITP-GB, Page 477; EA-AC65-9A, Page 465

5546. ANSWER #4
When an aircraft is sold, the Airworthiness Certificate is transferred with the aircraft.
REFERENCE: EA-ITP-GB, Page 546; EA-FAR 21.179

5547. ANSWER #1
If an Airworthiness Directive applies to a particular aircraft, the owner has no choice but to comply with it, or discontinue operating the aircraft.
REFERENCE: EA-FAR 39.3

5548. ANSWER #2
The issuance of Airworthiness Certificates is covered in FAR 21. FAR parts 23 and 25 are concerned with Airworthiness Standards, (not certification), and part 39 covers Airworthiness Directives.
REFERENCE: EA-FAR 21, Subpart H

5549. ANSWER #2
Older aircraft of which fewer than 50 remain in service, or any aircraft of which fewer than 50 were produced or remain in service, are found in Volume VI of the production design approvals. This volume is titled Aircraft Listing, Aircraft Engine and Propeller Listing.
REFERENCE: EA-ITP-GB, Page 474; Volume VI, Production Design Approvals

5550. ANSWER #4
All the specifications on a certificated propeller can be found in the propeller type certificate data sheet.
REFERENCE: EA-ITP-GB, Pages 483-484

5551. ANSWER #4
Of the items listed in this question, only the control surface movements would be found in the Type Certificate Data Sheet.
REFERENCE: EA-ITP-GB, Page 478; EA-AC65-9A, Page 466

5552. ANSWER #4
The placards required on an aircraft are identified in the Aircraft Specifications or Type Certificate Data Sheet.
REFERENCE: EA-ITP-GB, Page 479; EA-AC65-9A, Page 466

5553. ANSWER #1
Volume VI of the production design approvals contains specifications on aircraft of which fewer than 50 were produced or fewer than 50 remain in service. This volume is titled Aircraft Listings, Aircraft Engine and Propeller Listing.
REFERENCE: EA-ITP-GB, Page 474; Volume VI, Production Design Approvals

5554. ANSWER #4
Complying with the data in a Supplemental Type Certificate is a major alteration. Major alterations most definitely require approval from the person who performed the alteration and from the inspector who approves the aircraft for return to service. To install an item which is manufactured in accordance with the Technical Standard Order system does require approval when it is installed in aircraft.
REFERENCE: EA-FAR 21, Subpart E; EA-FAR 21, Subpart O

5555. ANSWER #4
FAR 23 covers the airworthiness standards for airplanes certificated in the normal, utility, and acrobatic categories. One of the standards identified in this regulation is the range markings on instruments.
REFERENCE: EA-FAR 23.1543

5556. ANSWER #2
Propellers most definitely are included in the Airworthiness Directive system. Airworthiness Directives are issued on all products when the need arises. Powerplant certified mechanics are allowed to perform minor repairs on aircraft propellers and return to service, provided they have the necessary experience, equipment, and technical data.
REFERENCE: EA-AC65-9A, Page 465; EA-FAR 65.87

5557. ANSWER #1
The aircraft Type Certificate Data Sheet contains information on the minimum octane rating of fuel to use and on the aircraft's fuel quantity.
REFERENCE: EA-ITP-GB, Pages 477-478; EA-AC65-9A, Pages 465-466

5558. ANSWER #2
Advisory Circular 43.13-2A provides acceptable methods of performing aircraft alteration, but it is not approved data. This is a recent interpretation by the FAA, because in the past this advisory circular was considered to be approved data.
REFERENCE: Recent ruling by the FAA. Advisory Circular 65-19C would indicate that ACs are approved data.

5559. ANSWER #2
Whenever an Airworthiness Directive is complied with, it must be recorded in the aircraft maintenance record.
REFERENCE: EA-AC65-19B, Chapter 9 (c), Appendix 1

5560. ANSWER #4
Statement #1 in this question is not true because the advisory circulars are included, and they are not approved data. Statement #2 is true because all the items listed are approved data.
REFERENCE: Recent ruling by the FAA. Advisory Circular 65-19C would indicate that ACs are approved data.

5561. ANSWER #4
The aircraft described in this question would fall under paragraph 1. Paragraph 1 applies to aircraft with less than 500 hours, and it states that after the AD is complied with that it should be repeated every 200 hours. On the aircraft in question, compliance first took place at 454 hours, and the aircraft now has 468 hours. This leaves 186 hours til the next compliance is due.
REFERENCE: EA-AC65-19B, Chapter 9(c)

5562. ANSWER #1
The nose door link shows a lubricant of oil, general purpose, low temperature lubricant.
REFERENCE: Testbook Figure 52

5563. ANSWER #1
The pivot bushings show a square frequency symbol, which identifies 100 hour intervals.
REFERENCE: Testbook Figure 52

5564. ANSWER #4
The nose door link shows a symbol for the squirt can as a method of applying the lubricant.
REFERENCE: Testbook Figure 52

5565. ANSWER #4
According to FAR 23, the flap operating range on an airspeed indicator is marked with a white arc.
REFERENCE: EA-FAR 23.1545(b)(4)

5566. ANSWER #1
The statement identifies that subsequent inspections will be made each second 50-hour period. Two 50-hour periods make the inspection intervals 100 hours.
REFERENCE: EA-AC65-19B, Chapter 9(c)

5567. ANSWER #1
The statement identifies that subsequent inspections will be made each third 50-hour period. Three 50-hour periods make the inspection intervals 150 hours.
REFERENCE: EA-AC65-19B, Chapter 9(c)

5568. ANSWER #4
FAR 65 authorizes powerplant mechanics to perform the 100-hour inspection on powerplants and propellers and, if appropriate, return the same to service.
REFERENCE: EA-FAR 65.87

5569. ANSWER #4
To approve the entire aircraft for return to service after a 100-hour inspection, the mechanic must have both the airframe and the powerplant ratings.
REFERENCE: EA-FAR 65.85 and 65.87

5570. ANSWER #2
Repairing, whether it be on the airframe or on the powerplant, can mean a variety of things. Because it does not state otherwise, we will assume that the question is referring to a minor repair. Minor repairs deal with the restoring of the airframe to a condition for safe operation after damage or deterioration.
REFERENCE: EA-FAR 43.13

5571. ANSWER #3
According to FAR 43, the replacement of fabric on fabric covered parts of aircraft is an airframe major repair.
REFERENCE: EA-FAR 43, Appendix A(b)(1)(xxvii)

5572. ANSWER #2
According to FAR 43, the repair of portions of skin sheets by making additional seams is an airframe major repair.
REFERENCE: EA-FAR 43, Appendix A(b)(1)(xxiii)

5573. ANSWER #3
According to FAR 65, an aircraft mechanic with the airframe and powerplant ratings is authorized to perform a 100-hour inspection and return the aircraft to service.
REFERENCE: EA-FAR 65.85 and 65.87

5574. ANSWER #3
FAR 65 authorizes certified A&P mechanics to perform the 100-hour inspection. It does not, however, authorize the mechanic to supervise another person doing the inspection. In FAR 43, the tasks which a supervised person are allowed to perform are spelled out, and the inspections required by FAR 91 are not allowed.
REFERENCE: EA-FAR 65.81 and 65.87; EA-FAR 43.3(d)

5575. ANSWER #4
According to FAR 65, a mechanics certificate is valid until surrendered, suspended, or revoked.
REFERENCE: EA-FAR 65.15

5576. ANSWER #4
When stating the privileges and limitations of a mechanic, FAR 65 specifically prohibits mechanics from repairing or altering instruments.
REFERENCE: EA-FAR 65.81

5577. ANSWER #1
Although powerplant certified mechanics are not allowed to perform major alterations or repairs to propellers, they are allowed to perform minor alterations and repairs.
REFERENCE: EA-FAR 65.81 and 65.87

5578. ANSWER #1
Powerplant certified mechanics are allowed to perform minor repairs to and minor alterations of propellers.
REFERENCE: EA-FAR 65.81 and 65.87

5579. ANSWER #4
FAA certified mechanics are authorized to perform minor alterations, appropriate to the rating(s) they hold, and return the item to service.
REFERENCE: EA-FAR 65.81

5580. ANSWER #2
FAR 65 authorizes the powerplant certified mechanic to perform the 100-hour inspection on the powerplant, propeller, and component parts thereof, and approve the items for return to service.
REFERENCE: EA-FAR 65.87

5581. ANSWER #4
To be considered current, a certified mechanic must, for at least six months out of twenty four, serve as a mechanic under his certificate and rating.
REFERENCE: EA-FAR 65.83

5582. ANSWER #2
FAR 65 imposes a limitation on certified mechanics when it states that they cannot perform or supervise major repairs to or major alterations of propellers. In addition, it states that certified mechanics cannot do any repair to or altering of instruments.
REFERENCE: EA-FAR 65.81

5583. ANSWER #2
In FAR 65 it states, that within 30 days after any change in permanent mailing address, the holder of a certificate issued under this Part (FAR 65) shall notify the FAA of his new address.
REFERENCE: EA-FAR 65.21

5584. ANSWER #1
Whenever maintenance is being performed on an aircraft, it is the responsiblity of the person doing the work to ensure that the parts being used meet the appropriate standards.
REFERENCE: EA-FAR 43.13(b)

5585. ANSWER #4
When replacing a part on an aircraft with a new part which is identical, a minor repair is being done (providing the part is not a primary structural member needing riveting or welding).
REFERENCE: EA-FAR 43, Appendix A(b)(1)

5586. ANSWER #1
A certified mechanic with the powerplant rating has the authority to perform a 100-hour inspection on the powerplant and propeller, and return the same to service.
REFERENCE: EA-FAR 65.87

5587. ANSWER #2
Any repair or alteration of instruments should be performed by an FAA approved instrument repair station.
REFERENCE: EA-AC43.13-1A, Page 303